QUANTITATIVE NEUROSCIENCE

KLUWER ACADEMIC PUBLISHERS
Boston / Dordrecht / New York / London

Biocomputing
Volume 2

QUANTITATIVE NEUROSCIENCE
Models, Algorithms, Diagnostics, and Therapeutic Applications

edited by

P. M Pardalos
Departments of Industrial and Systems Engineering, Biomedical Engineering, and McKnight Brain Institute
University of Florida.

J. C. Sackellares
Departments of Neuroscience, Neurology, Pediatrics, and Biomedical Engineering
University of Florida; VA Medical Center.

P. R. Carney
Departments of Pediatrics, Neurology, and Neuroscience
University of Florida.

L. D. Iasemidis
Department of Biomedical Engineering
Arizona State University.

KLUWER ACADEMIC PUBLISHERS
Boston / Dordrecht / New York / London

Distributors for North, Central and South America:
Kluwer Academic Publishers
101 Philip Drive
Assinippi Park
Norwell, Massachusetts 02061 USA
Telephone (781) 871-6600
Fax (781) 871-6528
E-Mail <kluwer@wkap.com>

Distributors for all other countries:
Kluwer Academic Publishers Group
Post Office Box 322
3300 AH Dordrecht, THE NETHERLANDS
Telephone 31 78 6576 000
Fax 31 78 6576 474
E-Mail <orderdept@wkap.nl>

Electronic Services <http://www.wkap.nl>

Library of Congress Cataloging-in-Publication

Pardalos, P. M./ Sackellares, J. C./ Carney, P. R./ Iasemidis, L. D.
Quantitative Neuroscience: Models, Algorithms, Diagnostics, and Therapeutic Applications
ISBN 1-4020-7751-3

Printed on acid-free paper.

Printed in the United States of America

Contents

Preface

Advances in the field of signal processing, nonlinear dynamics, statistics, and optimization theory, combined with marked improvement in instrumentation and development of computers systems, have made it possible to apply the power of mathematics to the task of understanding the human brain. This veritable revolution already has resulted in widespread availability of high resolution neuroimaging devices in clinical as well as research settings.

Breakthroughs in functional imaging are not far behind. Mathematical techniques developed for the study of complex nonlinear systems and chaos already are being used to explore the complex nonlinear dynamics of human brain physiology. Global optimization is being applied to data mining expeditions in an effort to find knowledge in the vast amount of information being generated by neuroimaging and neurophysiological investigations. These breakthroughs in the ability to obtain, store and analyze large datasets offer, for the first time, exciting opportunities to explore the mechanisms underlying normal brain function as well as the affects of diseases such as epilepsy, sleep disorders, movement disorders, and cognitive disorders that affect millions of people every year. Application of these powerful tools to the study of the human brain requires, by necessity, collaboration among scientists, engineers, neurobiologists and clinicians. Each discipline brings to the table unique knowledge, unique approaches to problem solving, and a unique language.

This book contains refereed invited papers submitted at the conference on *Quantitative Neurosciences: Models, Algorithms, Diagnosis, and Therapeutic Applications* held at the University of Florida, on February 5–7, 2003. The conference evolved from a growing awareness among acquaintances representing a diverse spectrum of scientists, of an increasing need for recurrent and ongoing evaluation of the rapidly accumulating knowledge of quantitative neuroscientific correlates. The success of the conference has been due to the remarkable ability of these scientists of markedly different persuasions to expedite and stimulate discussion, to facilitate exchange, to break down interdisciplinary barriers with unifying cross talk. This heterogeneous group included engineers, mathematicians, physicists, neurobiologists, neurologists, child neurologists, neurophysiologists, neuropsychologists, and students. The papers of these book reflect the topics discussed at the conference. A recurrent theme is evident in this book not only involving cerebral organization, but also its underlying neuroanatomical/neuroquantitative/neurophysiological substata, methods of investigation, algorithms, and potential therapeutic applications. The book is addressed to faculty, graduate students, and researchers in neurosciences, biomedical engineering, and statistics.

The editors want to take this opportunity to express gratitude to Ortho-McNeil Pharmaceuticals, Inc., University of Florida College of Medicine, University of Florida McKnight Brain Institute, University of Florida Child Research Institute, University of Florida College of Engineering, National Institutes of Health, and the University of Florida Division of Sponsored Research.

Many thanks go especially to Dr. Walter Freeman, Professor, University of California, Berkeley, Dr. Kirk Frey, Professor, University of Michigan, Ann Arbor, and Dr. John Milton, Professor, Department of Neurology, University of Chicago for moderating the final discussion. Special thanks and appreciation go to Ms. Danielle Becker who helped with the conference and Mr. Bruno H. Chiarini for assisting us in the the preparation of the camera ready Latex form of this book. Finally, we would like to thank Kluwer Academic Publishers for their assistance.

PANOS M. PARDALOS, J. CHRIS SACKELLARES, PAUL CARNEY, AND LEONIDAS IASEMIDIS

"Men ought to know that from the brain and from the brain only arise,
our pleasures, joys, laughter and sorrows...
Through it, we think, see, hear and distinguish the ugly from the beautiful,
the bad from the good, the pleasant from the unpleasant...
To consciousness the brain is messenger."

Hippocrates (c.460-377 BC)
On the Sacred Disease.

Chapter 1

APPLICATIONS OF GLOBAL OPTIMIZATION AND DYNAMICAL SYSTEMS TO PREDICTION OF EPILEPTIC SEIZURES*

W. Chaovalitwongse
Department of Industrial and Systems Engineering
University of Florida
arty@epilepsy.health.ufl.edu

P.M. Pardalos
Departments of Industrial and Systems Engineering, Computer Science, and Biomedical Engineering
University of Florida
pardalos@cao.ise.ufl.edu

L.D. Iasemidis
Department of Biomedical Engineering
Arizona State University
Leon.Iasemidis@asu.edu

D.-S. Shiau
Department of Neuroscience
University of Florida
shiau@epilepsy.health.ufl.edu

*Research was partially supported by the Medical Research Service of the Department of Veterans Affairs, grants from the Department of Veterans Affairs Research, the NSF grants DBI-980821, EIA-9872509, and NIH/NIBIB (8R01EB002089-03).

J.C. Sackellares
Departments of Neuroscience, Neurology, and Biomedical Engineering
University of Florida
The Malcom Randall Department of Veterans Affairs Medical Center, Gainesville
sackellares@epilepsy.health.ufl.edu

Abstract Seizure occurrences seem to be random and unpredictable. However, recent studies in epileptic patients suggest that seizures are deterministic rather than random. There is growing evidence that seizures develop minutes to hours before clinical onset. Our previous studies have shown that quantitative analysis based on chaos theory of long-term intracranial electroencephalogram (EEG) recordings may enable us to observe the seizure's development in advance before clinical onset. The period of seizure's development is called a preictal transition period, which is characterized by gradual dynamical changes in EEG signals of critical electrode sites from asymptomatic interictal state to seizure. Techniques used to detect a preictal transition include statistical analysis of EEG signals, optimization techniques, and nonlinear dynamics. In this paper, we herein present optimization techniques, specifically multi-quadratic 0-1 programming, for the selection of the cortical sites that are involved with seizure's development during the preictal transition period. The results of this study can be used as a criterion to pre-select the critical electrode sites that can be used to predict epileptic seizures.

Keywords: Multi-Quadratic 0-1 programming, Lyapunov exponents, EEG, Seizure prediction

1. Introduction

In the last decade, time series analysis based on chaos theory and the theory of nonlinear dynamics, which are among the most interesting and growing research topics, has been applied to time series data with some degree of success. The concepts of chaos theory and theory of nonlinear dynamics have not only been useful to analyze specific systems of ordinary differential equations or iterated maps, but have also offered new techniques for time series analysis. Moreover, a variety of experiments have shown that a recorded time series is driven by a deterministic dynamical system with a low dimensional chaotic attractor, which is defined as the phase space point or set of points representing the various possible steady-state conditions of a system; an equilibrium state or group of states to which a dynamical system converges. Thus, the theories of chaos and nonlinear dynamics have provided new theoretical and conceptual tools that allow us to capture, understand, and link the complex behaviors of simple systems together. Characterization and quantification of the dynamics of nonlinear time series are also important steps toward understanding the nature of random behavior and may enable us to predict the occurrences of some specific events which follow temporal dynamical patterns in the time series.

In this paper, we are concerned with the problem of predicting target episodes of events and discovering temporal patterns in multiple time series governing the related target episodes of events. Traditional linear and nonlinear time series analysis has been routinely used but did not seem to successfully give insight into the characteristic and mechanism of time series because these methods are limited by the stationary requirement of the time series and the normality and independence requirements of the residuals. These limitations and the lack of insight into the characteristic and mechanism of time series for making better event-predictions of traditional time series analysis are resolved by the development of new time series data mining concepts, which generalizes data mining concepts, dynamical approaches in chaos theory, and optimization techniques to the areas of time series analysis. These concepts are used to develop new techniques for the prediction of the time series arising in real world problems (e.g., electroencephalogram (EEG) time series) as well as to conduct advanced studies on the subject. The developed techniques use a combination of data mining techniques, dynamical approaches, and optimization techniques applied to time series data, with the objective of discovering temporal patterns in time series and then predicting events of interest. Specifically, this paper integrates methods based on chaos theory, statistical analysis and optimization techniques to identify complex (nonperiodic, nonlinear, irregular, and chaotic) characteristics and predict the onset of a target event from complex real world time series. In addition, we also focus on the statistical and optimization problems that enable us to detect statistically significant temporal patterns that can be used to characterize and predict the onset of target events in the times series. Identifying temporal patterns in multiple time series is combinatorial in nature, operating with the selection of critical components in the system of interest. Therefore, the optimization techniques are developed to improve the performance of prediction in the time series by identifying critical temporal patterns related to the target events (e.g., dynamical parameter settings and selecting of critical components). For instance, several alternative optimization methods for selecting the critical components in the systems are employed and a novel combination of methods for determining the optimal parameters can be applied to systems with one or more hidden variables, which can be used to reconstruct maps or differential equations of the dynamics of the system. Motivated by the spinning glass model, the problem of characterizing and identifying temporal patterns is ideally suited to 0-1 (two states) problems.

Herein, we direct our applications to bioengineering problems, particularly epilepsy and brain disorders. Epilepsy is among the most common disorders of the nervous system and consists of more than 40 clinical syndromes affecting 50 million people worldwide (approximately 1% of the population).

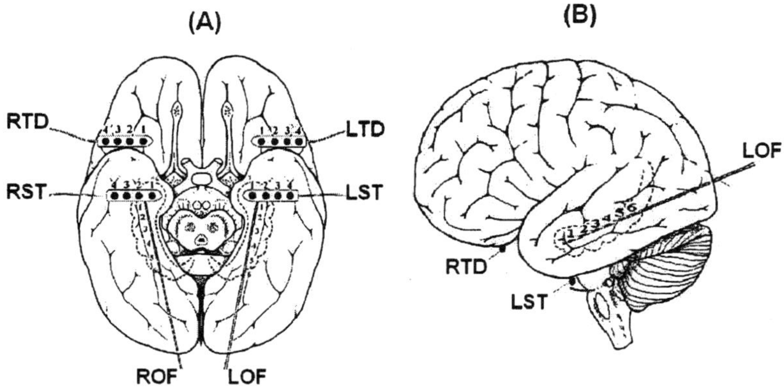

Figure 1.1. (A) Inferior transverse and (B) lateral views of the brain, illustrating approximate depth and subdural electrode placement for EEG recordings are depicted. Subdural electrode strips are placed over the left orbitofrontal (LOF), right orbitofrontal (ROF), left subtemporal (LST), and right subtemporal (RST) cortex. Depth electrodes are placed in the left temporal depth (LTD) and right temporal depth (RTD) to record hippocampal activity.

Epilepsy is characterized by intermittent seizures, that is, intermittent paroxysmal rhythmic electrical discharges within the cerebrum that disrupt normal brain function. Approximately 25 to 30% of patients receiving medication have inadequate seizure control. In other words, about 25% of patients with epilepsy have seizures that are resistant (refractory) to medical therapy. There is a localized structural change in neuronal circuitry within the cerebrum which produces organized quasi-rhythmic discharges in some types of epilepsy (i.e., focal or partial epilepsy). These discharges then spread from the region of origin (epileptogenic zone) to activate other areas of the cerebral hemisphere. Although the macroscopic and microscopic features of the epileptogenic zone have been comprehended, the mechanism by which these fixed disturbances in local circuitry produce intermittent disturbances of brain function cannot be explained and understood. The development of the epileptic state can be considered as changes in network circuitry of neurons in the brain. When neuronal networks are activated, they produce a change in voltage potential, which can be captured by an EEG. These changes are reflected by wriggling lines along the time axis in a typical EEG recording. A typical electrode montage for such recordings is shown in Figure 1.1. The EEG onset of a typical epileptic seizure of a focal origin recorded with this montage is illustrated in Figure 1.2. Figures 1.3 and 1.4 show the preictal state and postictal state of a typical epileptic seizure, respectively.

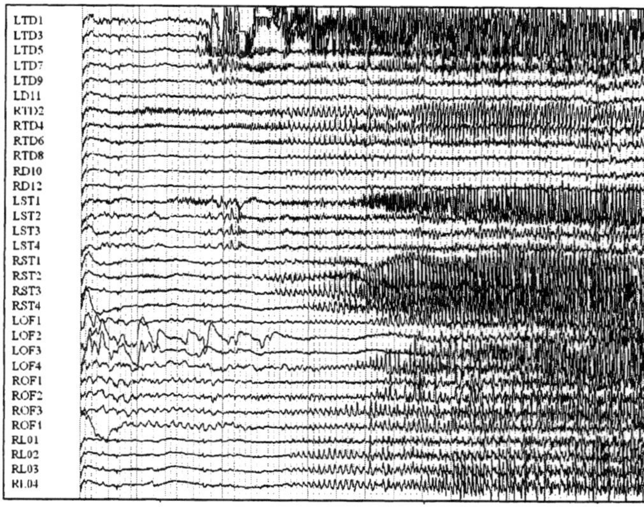

Figure 1.2. Twenty-second EEG recording of the onset of a typical epileptic seizure obtained from 32 electrodes. Each horizontal trace represents the voltage recorded from electrode sites listed in the left column (see Figure 1.1 for anatomical location of electrodes).

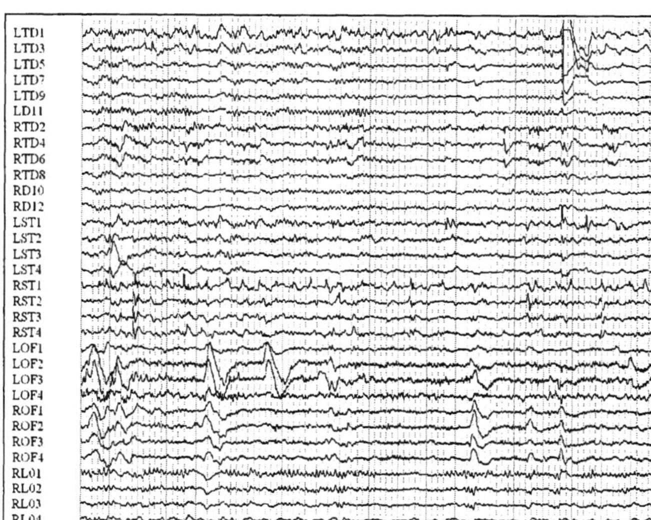

Figure 1.3. Twenty-second EEG recording of the preictal state of a typical epileptic seizure obtained from 32 electrodes. Each horizontal trace represents the voltage recorded from electrode sites listed in the left column (see Figure 1.1 for anatomical location of electrodes).

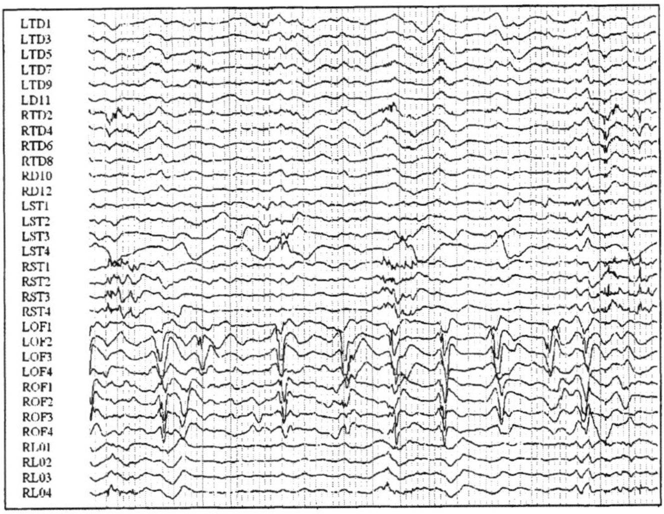

Figure 1.4. Twenty-second EEG recording of the postical state of a typical epileptic seizure obtained from 32 electrodes. Each horizontal trace represents the voltage recorded from electrode sites listed in the left column (see Figure 1.1 for anatomical location of electrodes).

We are specifically interested in the prediction of epileptic seizures, whose occurrence seems to be random and unpredictable. In essence, we integrate the developed techniques from data mining concepts, dynamical approaches in chaos theory, and optimization techniques as a set of tools used to extract dynamical changes in the EEG time series that precede a seizure. Specifically in this framework, studies based on chaos theory of the spatiotemporal dynamics in EEG's from patients with temporal lobe epilepsy demonstrate a pre-ictal transition (temporal patterns of dynamical changes in multiple EEG recordings), characterized by a progressive convergence (entrainment) of dynamical measures (e.g., short–term maximum Lyapunov exponents – STL_{max}) at specific anatomical areas in the neocortex and hippocampus before the seizure onset.

The organization of the succeeding sections of this chapter is as follows. The background and the method used to estimate STL_{max} and $\bar{\Omega}$ and the spatiotemporal dynamical analysis is described in section 2. In section 3, The multi-quadratic 0-1 programming for selection of critical cortical sites is addressed as well as the method to test the hypothesis and the results. The conclusions and discussed are addressed in the final section 5.

2. Background

In the last decade, several quantitative system approaches incorporating statistical techniques nonlinear methods based on chaos theory have been successfully used to study epilepsy because the aperiodic and unstable behavior of the epileptic brain is suitable to nonlinear techniques that allow precise tracking of the temporal evolution. Our previous studies have shown that seizures are deterministic rather than random. Consequently, studies of the spatiotemporal dynamics in long-term intracranial EEG's, from patients with temporal lobe epilepsy, demonstrated the predictability of epileptic seizures; that is seizures develop minutes to hours before clinical onset. The period of seizure's development is called a preictal transition period, which is characterized by gradual dynamical changes in EEG signals of critical electrode sites approximately $1/2$ to 1 hour duration before the ictal onset [9,11,13,15,23,24]. During a preictal transition period, gradual dynamical changes can be exposed by a progressive convergence (entrainment) of dynamical measures (e.g. short-term maximum Lyapunov exponents - STL_{max}) at specific anatomical areas and cortical sites, in the neocortex and hippocampus. Another measure we have used in the state space created from the EEG at individual electrode sites in the brain, average angular frequency ($\bar{\Omega}$), has produced promising results too. The value of $\bar{\Omega}$ quantifies the average rate of the temporal change in the state of a system and is measured in rads/sec. Although the existence of the preictal transition period has recently been confirmed and further defined by other investigators [4,5,16,17,21], the characterization of this spatiotemporal transition is still far from complete. For instance, even in the same patient, different set of cortical sites may exhibit preictal transition from one seizure to the next. In addition, this convergence of the normal sites with the epileptogenic focus (critical cortical sites) is reset after each seizure [14]. Therefore, complete or partial postictal resetting of preictal transition of the epileptic brain, affects the route to the subsequent seizure, contributing to the apparently non-stationary nature of the entrainment process. In those studies, however, the critical site selections are not trivial but extremely important since most groups of brain sites are irrelevant to the occurrences of the seizures and only certain groups of sites have dynamical convergence in the preictal transition.

Since the brain is a nonstationary system, algorithms used to estimate measures of the brain dynamics should be capable of automatically identifying and appropriately weighing existing transients in the data. In a chaotic system, orbits originating from similar initial conditions (nearby points in the state space) diverge exponentially (expansion process). The rate of divergence is an important aspect of the system dynamics and is reflected in the value of Lyapunov exponents and dynamical phase.

2.1. Estimation of Short Term Largest Lyapunov Exponents

The method we developed for estimation of Short Term Largest Lyapunov Exponents (STL_{max}), an estimate of L_{max} for nonstationary data, is explained in detail elsewhere [8,10,26]. Herein we will present only a short description of our method. Construction of the embedding phase space from a data segment x(t) of duration T is made with the method of delays. The vectors X_i in the phase space (see Figure 1.5) are constructed as:

$$X_i = (x(t_i), x(t_i + \tau) \ldots x(t_i + (p-1) * \tau)) \tag{1}$$

where τ is the selected time lag between the components of each vector in the phase space, p is the selected dimension of the embedding phase space, and $t_i \in [1, T - (p-1)\tau]$. If we denote by L the estimate of the short term largest Lyapunov exponent STL_{max} then:

$$L = \frac{1}{N_a \Delta t} \sum_{i=1}^{N_a} \log_2 \frac{|\delta X_{i,j}(\Delta t)|}{|\delta X_{i,j}(0)|} \tag{2}$$

with

$$\delta X_{i,j}(0) = X(t_i) - X(t_j) \tag{3}$$

$$\delta X_{i,j}(\Delta t) = X(t_i + \Delta t) - X(t_j + \Delta t) \tag{4}$$

where

- $X(t_i)$ is the point of the fiducial trajectory $\phi_t(X(t_0))$ with $t = t_i$, $X(t_0) = (x(t_0), \ldots, x(t_0 + (p-1) * \tau))$, and $X(t_j)$ is a properly chosen vector adjacent to $X(t_i)$ in the phase space (see below).

- $\delta X_{i,j}(0) = X(t_i) - X(t_j)$ is the displacement vector at t_i, that is, a perturbation of the fiducial orbit at t_i, and $\delta X_{i,j}(\Delta t) = X(t_i + \Delta t) - X(t_j + \Delta t)$ is the evolution of this perturbation after time Δt.

- $t_i = t_0 + (i-1) * \Delta t$ and $t_j = t_0 + (j-1) * \Delta t$, where $i \in [1, N_a]$ and $j \in [1, N]$ with $j \neq i$.

- Δt is the evolution time for $\delta X_{i,j}$, that is, the time one allows $\delta X_{i,j}$ to evolve in the phase space. If the evolution time Δt is given in sec, then L is in bits per second.

- t_0 is the initial time point of the fiducial trajectory and coincides with the time point of the first data in the data segment of analysis. In the estimation of L, for a complete scan of the attractor, t_0 should move within $[0, \Delta t]$.

- N_a is the number of local L_{max}'s that will be estimated within a duration T data segment. Therefore, if D_t is the sampling period of the time domain data, $T = (N-1)D_t = N_a \Delta t + (p-1)\tau$.

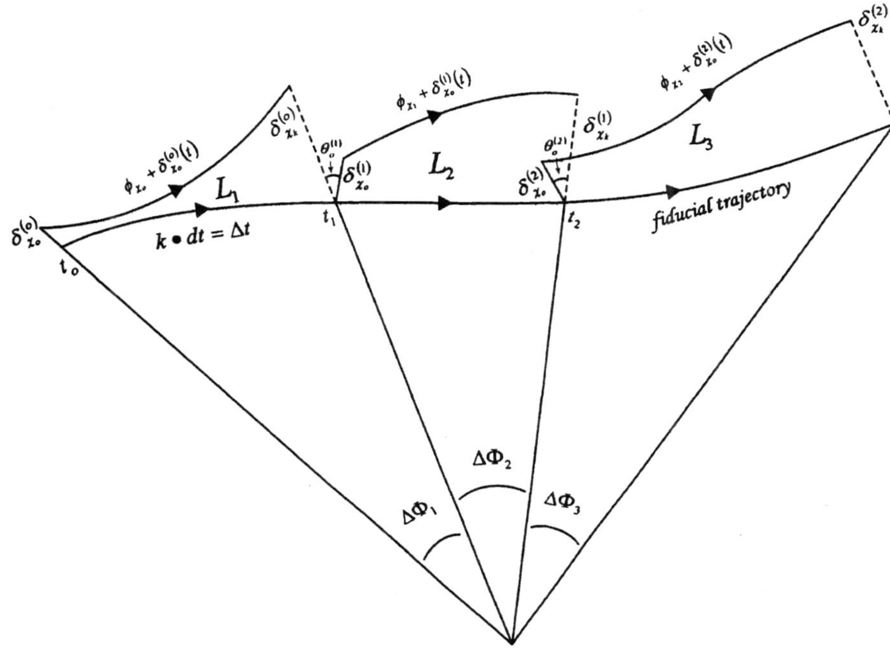

Figure 1.5. Diagram illustrating the estimation of STL_{max} measures in the state space. The fiducial trajectory, the first three local Lyapunov exponents (L_1, L_2, L_3), is shown

We computed the STL_{max} profiles using the method proposed by Iasemedis et al. [8], which is a modification of the method by Wolf et al. [26]. We call the measure short term to distinguish it from those used to study autonomous dynamical systems studies. Modification of the Wolf's algorithm is necessary to better estimate of STL_{max} in small data segments that include transients, such as interictal spikes. The modification is primarily in the searching procedure for a replacement vector at each point of a fiducial trajectory. For example, in our analysis of the EEG, we found that the crucial parameter of the L_{max} estimation procedure, in order to distinguish between the preictal, the ictal and the postictal stages, was not the evolution time Δt nor the angular separation $V_{i,j}$ between the evolved displacement vector $\delta X_{i-1,j}(\Delta t)$ and the candidate displacement vector $\delta X_{i,j}(0)$ (as it was claimed in Frank et al. [6]). The crucial parameter is the adaptive estimation in time and phase space of the magnitude bounds of the candidate displacement vector to avoid catastrophic replacements. Results from simulation data of known attractors have shown the improvement in the estimates of L achieved by using the proposed modifications [8]. In the preictal state, depicted in Figure 1.6, one can see a trend of STL_{max} toward lower values over the whole preictal period, with one prominent drop in the value of STL_{max} approximately 24 minutes prior to the seizure (denoted by an asterisk

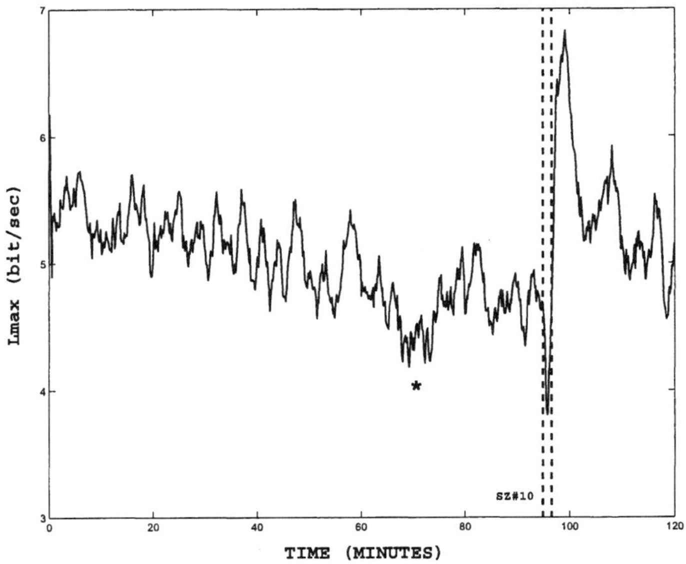

Figure 1.6. Smoothed STL_{max} profiles over 2 hours derived from an EEG signal recorded at RTD2 (patient 1). A seizure (SZ ♯ 10) started and ended between the two vertical dashed lines. The estimation of the L_{max} values was made by dividing the signal into non-overlapping segments of 10.24 sec each, using $p = 7$ and $\tau = 20$ msec for the phase space reconstruction. The smoothing was performed by a 10 point (1.6 minutes) moving average window over the generated STL_{max} profiles.

in the figure). This preictal drop in STL_{max} can be explained as an attempt of the system toward a new state of less degrees of freedom long before the actual seizure [11].

2.2. Estimation of Dynamical Phase (Angular Frequency)

Motivated by the representation of a state as a vector in the state space, we have defined the difference in phase between two evolved states $X(t_i)$ and $X(t_i + \Delta t)$ as $\Delta\Phi_i$ [12]. Then, denoting with $(\Delta\Phi)$ the average of the local phase differences $\Delta\Phi_i$ between the vectors in the state space, we have:

$$\Delta\Phi = \frac{1}{N_\alpha} \cdot \sum_{i=1}^{N_\alpha} \Delta\Phi_i \tag{5}$$

where N_α is the total number of phase differences estimated from the evolution of $X(t_i)$ to $X(t_i + \Delta t)$ in the state space, according to:

$$\Delta\Phi_i = \left| \arccos \frac{X(t_i) \cdot X(t_i + \Delta t)}{\| X(t_i) \| \cdot \| X(t_i + \Delta t) \|} \right| . \tag{6}$$

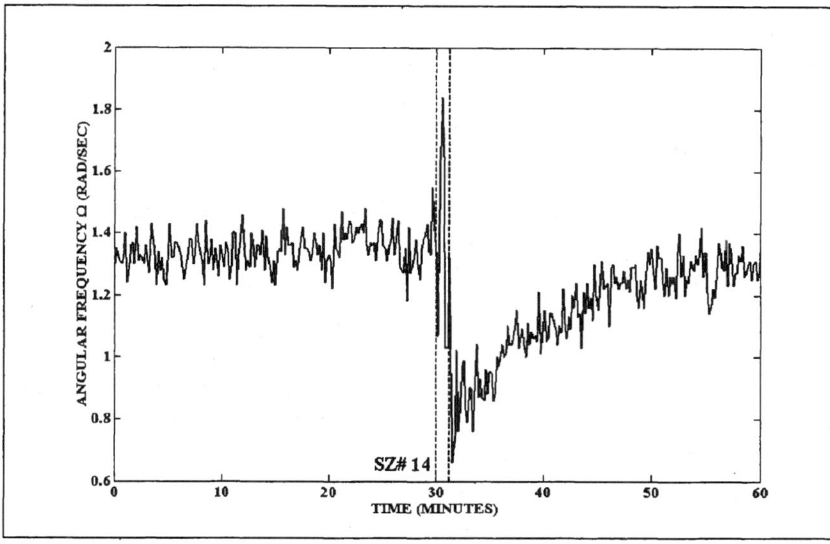

Figure 1.7. A typical $\bar{\Omega}$ profile before, during and after an epileptic seizure, estimated from the EEG recorded from a site in the epileptogenic hippocampus; the seizure occurred between the vertical lines.

Then, the average angular frequency $\bar{\Omega}$ is:

$$\bar{\Omega} = \frac{1}{\Delta t} \cdot \Delta\Phi. \tag{7}$$

If Δt is given in sec, then $\bar{\Omega}$ is given in rad/sec. Thus, while STL_{max} measures the local stability of the state of the system on average, $\bar{\Omega}$ measures how fast a local state of the system changes on average (e.g. dividing $\bar{\Omega}$ by 2π, the rate of the change of the state of the system is expressed in $sec^{-1} = Hz$).

An example of a typical $\bar{\Omega}$ profile over time is given in Figure 1.7. The values are estimated from a 60-minute-long EEG sample recorded from an electrode located in the epileptogenic hippocampus. The EEG sample includes a 2-minute seizure that occurs in the middle of the recording. The state space was reconstructed from sequential, non-overlapping EEG data segments of 2048 points (sampling frequency 200 Hz, hence each segment of 10.24 sec in duration) with $p = 7$ and $\tau = 4$, as for the estimation of STL_{max} profiles [12]. The preictal, ictal and postictal states correspond to medium, high and lower values of $\bar{\Omega}$ respectively. The highest $\bar{\Omega}$ values were observed during the ictal period, and higher $\bar{\Omega}$ values were observed during the preictal period than during the postictal period. This pattern roughly corresponds to the typical observation of higher frequencies in the original EEG signal ictally, and lower EEG frequencies postictally. However, these observations can hardly denote a long-term warning of an impending seizure.

2.3. Spatiotemporal Dynamical Analysis

Although a great deal is now known about low dimensional chaos, the erratic motion of dynamical systems described by a few variables, is understood about systems where the number of chaotic degrees of freedom becomes very large. Typically such systems show disorder in both space and time and are said to exhibit spatiotemporal chaos. Spatiotemporal chaos occurs when the system of coupled dynamical systems gives rise to dynamical behavior that exhibits both spatial disorder (as in rapid decay of spatial correlations) and temporal disorder (as in nonzero Lyapunov exponents). This is an extremely active, and rather unsettled area of research. The system under consideration (brain) has a spatial extent and, as such, information about the transition of the system towards the ictal state should also be included in the interactions of its spatial components. The preictal transition, progressive convergence of STL_{max} profiles, is another evidence of spatiotemporal chaos in the brain (shown in Figure 1.8). Having estimated the STL_{max} temporal profiles at individual cortical site, and as the brain proceeds towards the ictal state, the temporal evolution of the stability of each cortical site is quantified. The spatial dynamics of this transition are captured by considering relationship of the STL_{max} between different cortical sites. For example, if a similar transition occurs at different cortical sites, the STL_{max} of the involved sites are expected to converge to similar values prior to the transition. We have called such participating sites "critical sites", and such a convergence "dynamical entrainment." More specifically, in order for the dynamical entrainment to have a statistical content, we have allowed a period over which the mean of the differences of the STL_{max} values at two sites is estimated. We have used 60 STL_{max} values (i.e. moving windows of approximately of 10 minutes at each electrode site) to test the dynamical entrainment at the 0.01 statistical significance level. We employ the T-index as a measure of dynamical entrainment of STL_{max} profiles over time. The T-index at time t between electrode sites i and j is defined as:

$$T_{i,j}(t) = \sqrt{N} \times |E\{STL_{max,i} - STL_{max,j}\}|/\sigma_{i,j}(t) \tag{8}$$

where $E\{\cdot\}$ is the sample average difference for the $STL_{max,i} - STL_{max,j}$ estimated over a moving window $w_t(\lambda)$ defined as:

$$w_t(\lambda) = \begin{cases} 1 & \text{if } \lambda \in [t-N-1, t] \\ 0 & \text{if } \lambda \notin [t-N-1, t], \end{cases}$$

where N is the length of the moving window. Then, $\sigma_{i,j}(t)$ is the sample standard deviation of the STL_{max} differences between electrode sites i and j within the moving window $w_t(\lambda)$. The thus defined T-index follows a t-distribution with N-1 degrees of freedom. For the estimation of the $T_{i,j}(t)$ indices in our data we used $N = 60$ (i.e., average of 60 differences of STL_{max}

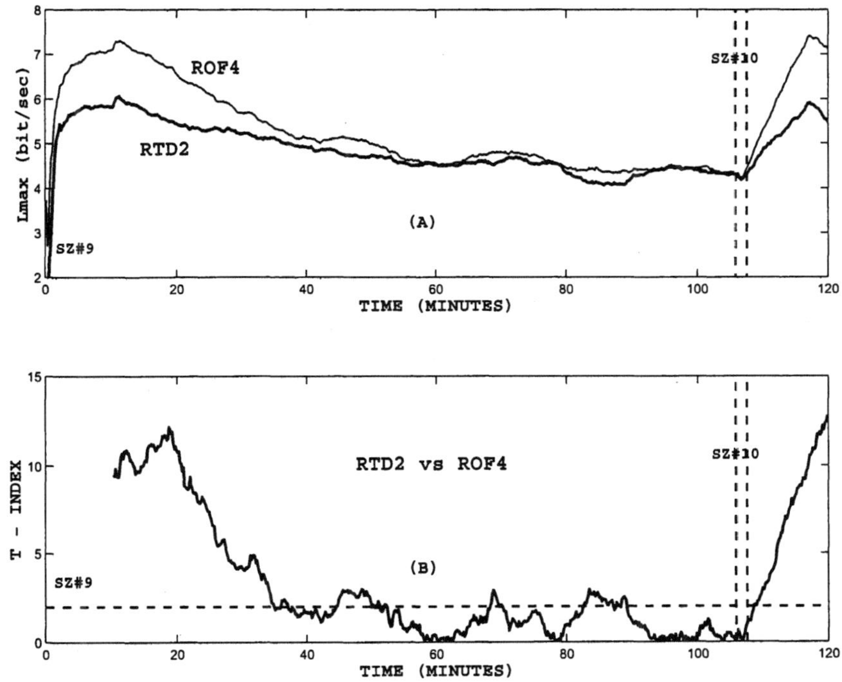

Figure 1.8. Dynamical entrainment of a pair of brain sites between seizure ♯ 9 and ♯ 10 (patient 1). (A) STL_{max} values at the normal site ROF4 approach the one at the epileptogenic site RTD2 about 35 minutes into the recording. Then, for 70 minutes up to seizure, the two sites interact with each other. (B) The T-index profile generated from the STL_{max} profiles in (A) is shown. The $\alpha = 0.05$ statistical significance level entrainment zone is also shown (dashed horizontal line). Entrainment of the two sites occurs when T-index values are within the depicted entrainment zone.

exponents between sites i and j per moving window of approximately 10 minute duration). Therefore, a two-sided t-test with $N - 1(= 59)$ degrees of freedom, at a statistical significance level α should be used to test the null hypothesis, H_o: "brain sites i and j acquire identical STL_{max} values at time t." In this experiment, we set $\alpha = 0.01$, the probability of a type I error, or better, the probability of falsely rejecting H_o if H_o is true, is 1%. For the T-index to pass this test, the $T_{i,j}(t)$ value should be within the interval [0,2.662]. In Figure 1.8(A), preictal entrainment (long before the occurrence of a seizure) and postictal disentrainment (after the occurrence of the seizure) of the STL_{max} profiles at two brain sites is shown. In Figure 1.8(B), this behavior is quantified by the T-index profile between these sites. From this figure, it is clear that attempts for a spatiotemporal entrainment between brain sites occur long before an epileptic seizure (first attempt about 70 minutes prior to seizure). Postictally, this entrainment is reset.

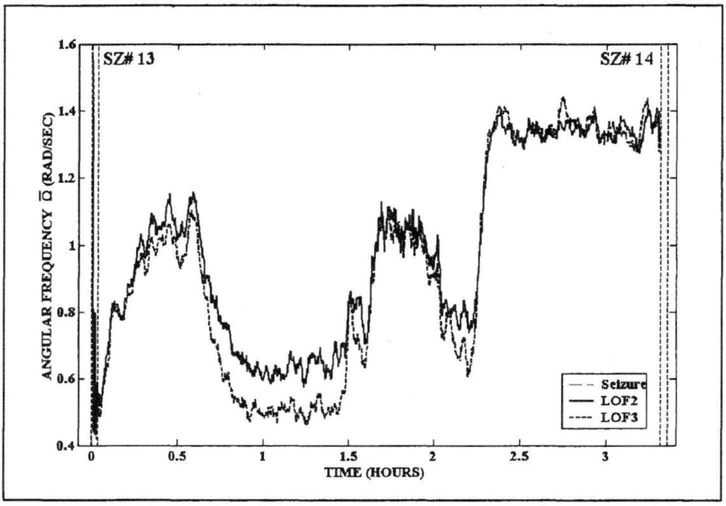

Figure 1.9. Angular frequency $\bar{\Omega}$ profiles from two left orbitofrontal electrode sites over 3.5 hours between seizures 13 and 14 (patient 1). The ictal periods of the two seizures are denoted by vertical lines

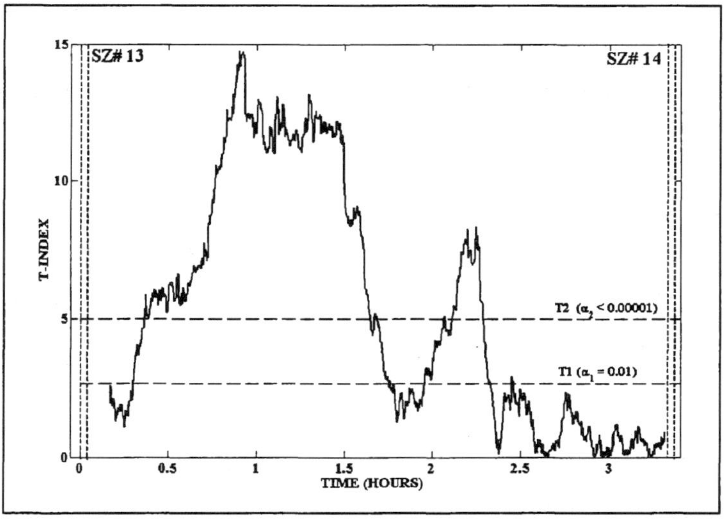

Figure 1.10. The T-index profile between two electrode sites whose $\bar{\Omega}$ profiles are depicted in Figure 1.9. The two sites are dynamically entrained 1.75 to 1.5 hours, as well as 1.2 hour prior to seizure's 14 onset. The T_1 and T_2 statistical thresholds are represented by the two horizontal lines.

In Figure 1.9, the $\bar{\Omega}$ profiles at two cortical sites are shown for the interval between seizures 13 and 14 in patient 1. For these cortical sites, a remarkable feature is observed: a long-term convergence of their $\bar{\Omega}$ profiles prior to seizure 14. We have called this convergence "dynamical entrainment" and we have quantified it by a T-statistic that provides a comparison between the two electrode sites (shown in Figure 1.10).

3. Method and Results

In this paper, we developed optimization techniques for electrode site selection to test the hypothesis that the set of cortical sites that participate in the preictal transition before the current seizure is most likely to participate in the preictal transition again before the next seizure. The set of participating sites is defined as a set of cortical sites with minimum difference in STL_{max} (most converged) prior to the current seizure and then reset after the seizure. To select the participating sites, we formulate this problem as a multi-quadratic 0-1 problem. We tested this hypothesis on the continuous long-term (3 to 12 days) multichannel intracranial EEG recordings that had been acquired from 3 patients with medically intractable temporal lobe epilepsy. The recordings were obtained as part of a pre-surgical clinical evaluation. They had been obtained using a Nicolet BMSI 4000 and 5000 recording systems, using a 0.1 Hz high-pass and a 70 Hz low-pass filter. Each record included a total of 28 to 32 intracranial electrodes (8 subdural and 6 hippocampal depth electrodes for each cerebral hemisphere).

In this framework, we first estimate STL_{max}, which measure of the order or disorder of EEG signals recorded from individual electrode sites. Based on STL_{max}, a multi-quadratic 0-1 programming problem is solved to identify participating electrode sites. The probabilities of participating in the next seizure preictal transition period of electrode sites selected from optimization problem and randomly selected electrode sites are then compared. The results of this study can be used as a criterion to pre-select the critical electrode sites that can be used to detect the preictal transition before an impending seizure.

4. Quadratic 0-1 Programming

In this paper we refer to the Sherrington-Kirkpatric Hamiltonian that describes the mean-field theory of the spin glasses where elements are placed on the vertices of a regular lattice, the magnetic interactions hold only for nearest neighbors and every element has only two states (Ising spin glasses [1–3,7,18]). One of the most interesting problems about this model is the determination of the minimal energy states (GROUND STATE problem).

For many years the Ising model has been a powerful tool in studying phase transitions in statistical physics. Such an Ising model can be described by a graph $G(V, E)$ having n vertices $\{v_1, \ldots, v_n\}$ and each edge $(i, j) \in E$ having a weight (interaction energy) J_{ij}. Each vertex v_i has a magnetic spin variable $\sigma_i \in \{-1, +1\}$ associated with it. An optimal spin configuration of minimum energy is obtained by minimizing the Hamiltonian

$$H(\sigma) = - \sum_{1 \leq i \leq j \leq n} J_{ij}\sigma_i\sigma_j \text{ over all } \sigma \in \{-1, +1\}^n.$$

This problem is equivalent to the combinatorial problem of quadratic bivalent programming [7].

Quadratic zero-one programming has been extensively used to study Ising spin glass models. This has motivated us to use quadratic 0-1 programming to select the critical cortical sites, where each electrode has only two states, and to determine the minimal-average T-index state. We formulated this problem as a quadratic 0-1 knapsack problem with objective function to minimize the average T-index (a measure of statistical distance between the mean values of STL_{max}) among electrode sites and the knapsack constraint to identify the number of critical cortical sites.

Let A be $n \times n$ matrix, whose each element $a_{i,j}$ represents the T-index between electrode i and j within 10-minute window before the onset of a seizure. Define $x = (x_1, ..., x_n)$, where each x_i represents the cortical electrode site i. If the cortical site i is selected to be one of the critical electrode sites, then $x_i = 1$; otherwise, $x_i = 0$.

A quadratic function is defined on R^n by

$$\min f(x) = x^T Ax, \text{ s.t. } x_i \in \{0, 1\}, i = 1, ..., n \tag{9}$$

where A is an $n \times n$ matrix [19, 20]. Throughout this section the following notations will be used.

- $\{0, 1\}^n$: set of n dimensional 0-1 vectors.
- $R^{n \times n}$: set of $n \times n$ dimensional real matrices.
- R^n: set of n dimensional real vectors.

Next, we add a linear constraint, $\sum_{i=1}^n x_i = k$, where k is the number of critical electrode sites that we want to select. We now consider the following linearly constrained quadratic 0-1 problem:

$$\bar{P} : \min f(x) = x^T Ax, \text{ s.t. } \sum_{i=1}^{n} x_i = k \text{ for some } k, x \in \{0, 1\}^n, A \in R^{n \times n}. \tag{10}$$

Problem $\bar{P}$ can be formulated as a quadratic 0-1 problem of the form as in (8) by using an exact penalty. If $A = (a_{ij})$ then let $M = 2[\sum_{j=1}^n \sum_{i=1}^n |a_{ij}|] + 1$.

Then, we have the following equivalent problem P as follows:

$$P : \min g(x) = x^T Ax + M(\sum_{i=1}^{n} x_i - k)^2, \text{ s.t. } x \in \{0,1\}^n, A \in R^{n \times n}. \quad (11)$$

To solve this problem, we considered 3 computational approaches. In the first approach, we solved (11) by applying a branch and bound algorithm with a dynamic rule for fixing variables [19, 20]. In the second approach, we use a linearization technique to formulate the quadratic integer programming (QIP) problem in (10) as an integer programming (IP) problem by introducing a new variable for each product of two variables and adding some additional constraints, and then formulated this problem as a linear 0-1 problem. In the third approach, we employed the Karush-Khun Tucker optimality conditions of the linearly constrained quadratic 0-1 problem in (10) to formulate this problem as a mixed-integer linear programming (MILP) problem. Details of the first approach can be found in [19, 20]; next, we discuss the second and the third approaches.

4.1. Conventional Linearization Approach

For each product $x_i x_j$, we introduce a new 0-1 variable, $x_{ij} = x_i x_j$ $(i \neq j)$. Note that $x_{ii} = x_i^2 = x_i$ for $x_i \in \{0, 1\}$. After linearization, the equivalent IP formulation is given by:

$$\min \sum_i \sum_j a_{ij} \; x_{ij} \quad (12)$$

$$\text{s.t.} \sum_{i=1}^{n} x_i = k, \quad (13)$$

$$x_{ij} \leq x_i, \text{ for } i, j = 1, ..., n \; (i \neq j) \quad (14)$$

$$x_{ij} \leq x_j, \text{ for } i, j = 1, ..., n \; (i \neq j) \quad (15)$$

$$x_i + x_j - 1 \leq x_{ij}, \text{ for } i, j = 1, ..., n \; (i \neq j) \quad (16)$$

where $x_i \in \{0, 1\}$ and $x_{ij} \in \{0, 1\}$, $i, j = 1, ..., n$.

The number of 0-1 variables has been increased to $O(n^2)$. Although, we can apply CPLEX 7.0 to solve problems with $n = 30$, this approach becomes computationally inefficient as n increases. Future technology will require the ability to efficiently solve problems with much larger values for n. For instance, micro-electrodes will be implanted in the future ($n > 1000$).

4.2. KKT Conditions Linearization Approach

Consider a linearly constrained quadratic problem given by:

$$\min z(x) = x^T Ax, \text{ s.t. } \sum_{i=1}^{n} x_i = k, x_i \geq 0, i = 1, ..., n. \quad (17)$$

We then have the following Karush-Kuhn Tucker conditions:

$$\begin{aligned} -2Ax + u.e + y &= 0 & (18) \\ \sum_{i=1}^{n} x_i &= k & (19) \\ y^T x &= 0, & (20) \end{aligned}$$

where u and y are Lagrangian multipliers. Note that u is a scalar and y is a column vector. We add the slack variables w, which is a column vector, to (18) and then denote a column vector $s = u.e + w$. We then have the KKT conditions given by:

$$\begin{aligned} -2Ax + y + s &= 0 \\ \sum_{i=1}^{n} x_i &= k \\ y^T x &= 0. \end{aligned}$$

We can formulate the above KKT conditions as a MILP formulation. The objective function is to minimize the summation of variables, s_i. Because x_i are 0-1 variables, we can replace the last constraint with $y_i \leq \mu(1 - x_i)$, for $i = 1, ..., n$, where $\mu \max_i \sum_{j=1}^{n} a_{ij} = \|A\|_\infty$. We then have the MILP formulation given by:

$$\begin{aligned} \min \sum_{i=1}^{n} & \quad s_i \\ \text{s.t. } -\sum_{j=1}^{n} a_{ij} x_j + s_i + y_i &= 0, \text{for } i = 1, ..., n \\ \sum_{i=1}^{n} x_i - k &= 0 & (21) \\ y_i - \mu(1 - x_i) &\leq 0, \text{for } i = 1, ..., n \end{aligned}$$

where $x_i \in \{0, 1\}$ and $s_i, y_i \geq 0$, for $i = 1, ..., n$.

Next, we want to show the optimality proof of the KKT conditions linearization approach. Consider a quadratic zero-one programming problem, which has the form

$$\min f(x) = x^T Ax, \text{ s.t. } Bx \geq b, \tag{22}$$

where A is an $n \times n$ matrix, whose each element $a_{ij} \geq 0,\ i, j = 1, \ldots, n$, $x \in \{0,1\}^n$, B is an $m \times n$ matrix, b is a constant vector, m and n are some integer numbers.

Let e be a vector of all 1's, i.e. $e = (1, \ldots, 1)^T$. Consider the following two problems:

P_1 : $\min f(x) = x^T Ax$, s.t. $Bx \geq b,\ x \in \{0,1\}^n$.

$\bar{P}_1$: $\min g(s) = e^T s$, s.t. $Ax - y - s = 0,\ Bx \geq b,\ y^T x = 0,\ x \in \{0,1\}^n,\ y_i \geq 0,\ s_i \geq 0$.

Let us prove the following theorem.

Theorem 1 *P_1 has an optimal solution x^0 iff there exist y^0, s^0 such that (x^0, y^0, s^0) is an optimal solution of $\bar{P}_1$.*

Proof 1 *Necessity. Let x^0 be an optimal solution of the problem P_1. Since all elements of the matrix A are nonnegative, it is obvious that $\exists y, s : y \geq 0,\ s \geq 0$ such that*

$$Ax^0 - y - s = 0, \tag{23}$$

$$y^T x^0 = 0. \tag{24}$$

Choose y^0, s^0 from the above defined set of y and s such that $e^T s^0$ is minimized. Then, we prove that (x^0, y^0, s^0) is an optimal solution of the problem $\bar{P}_1$.

Multiplying (23) by $(x^0)^T$, we obtain $(x^0)^T Ax^0 - (x^0)^T y^0 - (x^0)^T s^0 = 0$. Note from (24) that $(x^0)^T y^0 = 0$. Hence, we have

$$(x^0)^T Ax^0 = (x^0)^T s^0. \tag{25}$$

If we can prove that

$$(x^0)^T s^0 = e^T s^0, \tag{26}$$

then (x^0, y^0, s^0) is an optimal solution of $\bar{P}_1$. To prove that (26) holds, it is sufficient to show that, for any i if $x_i^0 = 0$ then $s_i^0 = 0$. We can prove this by contradiction.

Assume that for some i, $x_i^0 = 0$ and $s_i^0 > 0$, where (y^0, s^0) were chosen to minimize $e^T s^0$. Define vectors $\tilde{y}$ and $\tilde{s}$ as $\tilde{y}_i = y_i^0 + s_i^0$, $\tilde{s}_i = 0$ and for $i \neq j$ $\tilde{y}_j = y_j$, $\tilde{s}_j = s_j$. It easy to check that $(x^0,\ \tilde{y},\ \tilde{s})$ also satisfies (23), (24), and $e^T \tilde{s} < e^T s^0$. This contradicts with the initial assumption that s^0 and y^0 were chosen to minimize $e^T s^0$.

Sufficiency. The proof is similar.

It is easy to see from the complementarity constraint $y^T x = 0$ that for every i, where $x_i = 1$, we need to have $y_i = 0$; for every i, where $x_i = 0$, the value of y_i does not depend on this constraint. Also note from (23) that the value of y_i is upper bounded by the value of $M = \max_i \sum_{j=1}^{n} |a_{ij}| = \|A\|_\infty$. Therefore, we can reformulate $\bar{P}_1$ as a linear mixed 0–1 programming replacing the complementarity constraint $y^T x = 0$ by a linear constraint $y \leq M(e - x)$. As a result we obtain the following formulation:

$\tilde{P}_1$: $\min g(s) = e^T s$, s.t. $Ax - y - s = 0,\ Bx \geq b,\ y \leq M(e - x),\ s_i \geq 0,\ y_i \geq 0,\ x \in \{0,1\}^n$.

From Theorem 1, we have shown that problems P_1, $\bar{P}_1$, and $\tilde{P}_1$ are "equivalent." Therefore, QIP formulation in (10) is equivalent to the MILP formulation in (21). From (8), $T_{i,j}(t) = \sqrt{N} \times |E\{STL_{max,i} - STL_{max,j}\}|/\sigma_{i,j}(t)$, we note that every element in T-index matrix A is positive. For this reason, in every instance, by solving the MILP problem in (21) we can find the global solution to the original QIP problem in (10). Applying CPLEX 7.0, this problem can be easily solved with $n = 30$. In addition, this formulation is computationally efficient as n increases because the number of 0-1 variables is $O(n)$. From computational experiments, the above linear mixed integer 0-1 problem is the most efficient approach in our application, see Table 1.1 and Figure 1.11.

Table 1.1. Performance characteristics of two proposed approaches compared with complete enumerations

Number of selected electrodes	*KKT Conditions Approach*	*Linearization Approach*	*Complete Enumerations*
5 (out of 30)	297	656	15
6 (out of 30)	406	735	78
7 (out of 30)	609	968	313
8 (out of 30)	1797	2610	1141
9 (out of 30)	2562	5235	3578

4.3. Multi-Quadratic 0-1 Programming

Our group has shown dynamical resetting of the brain following seizures [14, 22, 25], that is, divergence of STL_{max} profiles after seizures. Therefore, we want to incorporate this finding with our existing critical electrode selection problem (QIP problem in (10)). Thus, we have to ensure that the optimal group of critical sites shows this divergence by adding one more quadratic constraint to the QIP problem in (10). The multi-quadratic integer programming (MQIP)

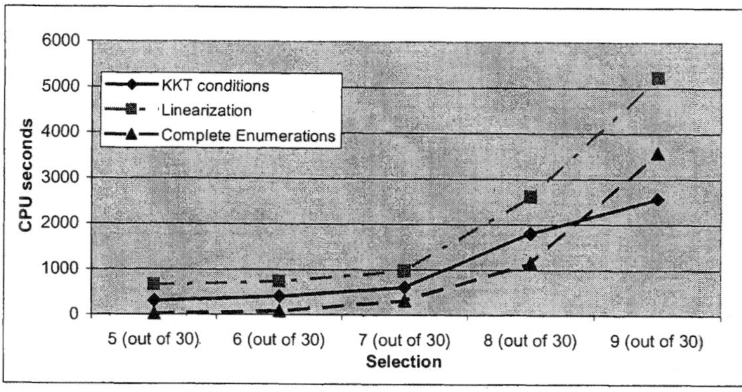

Figure 1.11. Performance characteristics of two proposed approaches compared with complete enumerations

problem is given by:

$$
\begin{array}{lll}
\min & x^T A x & \\
\text{s.t.} & \sum_{i=1}^{n} x_i & = k \\
 & x^T B x & \geq T_\alpha k(k-1)
\end{array}
\tag{27}
$$

where $x_i \in \{0,1\}\ \forall i \in \{1,...,n\}$.

Let B be $n \times n$ matrix, whose each element $b_{i,j}$ represents the T-index between electrode i and j within 10-minute window after the onset of a seizure. Note that the matrix $A = (a_{ij})$ is the T-index matrix of brain sites i and j within 10-minute windows before the onset of a seizure. T_α is the critical value of T-index, as previously defined, to reject H_o: "two brain sites acquire identical STL_{max} values within time window $w_t(\lambda)$."

With one more quadratic constraint, the problem in (27) becomes much harder to solve. Note that in the first approach, a branch and bound algorithm with a dynamic rule for fixing variables cannot be applied to solve this problem because of the additional quadratic constraint. However, we can modify the MIP formulation in (21) from the previous section and reformulate this problem by adding one more linearized constraint. The equivalent IP formulation is given by:

$$\begin{aligned}\min \sum_i \sum_j \quad & a_{ij} \quad x_{ij} \\ \text{s.t.} \sum_{i=1}^{n} x_i &= k, \\ x_{ij} &\le x_i, \text{ for } i,j = 1,...,n \ (i \neq j) \\ x_{ij} &\le x_j, \text{ for } i,j = 1,...,n \ (i \neq j) \\ x_i + x_j - 1 &\le x_{ij}, \text{ for } i,j = 1,...,n \ (i \neq j) \\ \sum_i \sum_j b_{ij} x_{ij} &\ge T_\alpha k(k-1)\end{aligned}$$

where $x_i \in \{0,1\}$ and $x_{ij} \in \{0,1\}$, $i,j = 1,...,n$. As we mentioned in the previous section, the above formulation is not computationally efficient as n increases.

Next, we want to show the optimality proof of the KKT conditions linearization approach, which can solve MQIP problem in (27) optimally. Now consider the case when we have a quadratic constraint. Let C be an $n \times n$ matrix, whose each element $c_{ij} \ge 0,\ i,j = 1,\ldots,n$.

Consider the following two problems:

P_2 : $\min f(x) = x^T Ax$, s.t. $Bx \ge b,\ x^T Cx \ge \alpha,\ x \in \{0,1\}^n$, α is a positive constant.

$\bar{P}_2$: $\min g(s) = e^T s$, s.t. $Ax - y - s = 0,\ Bx \ge b,\ y \le M(e-x),\ Cx - z \ge 0,\ e^T z \ge \alpha,\ z \le M'x,\ x \in \{0,1\}^n,\ y_i,\ s_i,\ z_i \ge 0$, where $M' = \|C\|_\infty$ and $M = \|A\|_\infty$.

Let us prove the following theorem.

Theorem 2 *P_2 has an optimal solution x^0 iff there exist $y^0,\ s^0,\ z^0$ such that (x^0, y^0, s^0, z^0) is an optimal solution of $\bar{P}_2$.*

Proof 2 *Necessity. From the proof of Theorem 1, it is obvious that we only need to show that if x^0 is an optimal solution of the problem P_2 then there exists vector z^0 such that every component is nonnegative, i.e. $z_i^0 \ge 0$, and the following constraints are satisfied:*

$$Cx^0 - z^0 \ge 0, \tag{28}$$

$$e^T z^0 \ge \alpha, \tag{29}$$

$$z^0 \le M'x^0. \tag{30}$$

From (30), note that if $x_i^0 = 0$ then we must have $z_i^0 = 0$. Similar to the proof of Theorem 1, we have that

$$e^T z^0 = (x^0)^T z^0. \tag{31}$$

Since z_i^0 is a real number and every element of the matrix C is nonnegative, then for all i, where we have $x_i^0 = 1$, we can choose $z_i^0 \geq 0$ such that $(Cx^0)_i = z_i^0$. Therefore, (28) and (30) are satisfied.

Multiplying (28) by $(x^0)^T$, from (31) we obtain that

$$(x^0)^T Cx^0 = (x^0)^T z^0 = e^T z^0 \tag{32}$$

and as x^0 is an optimal solution of the problem P_2 then (29) is satisfied:

$$e^T z^0 = (x^0)^T Cx^0 \geq \alpha \tag{33}$$

Sufficiency. The proof is similar.

From Theorem 2, we have shown that solving the MILP in (21) gives us the optimal solution, which is the global solution to the QIP problem in (10). In that proof, we show that we can also solve the MQIP problem in (27) by solving the following MILP formulation.

$$\min \sum_{i=1}^{n} s_i \tag{34}$$

$$\text{s.t.} \sum_{i=1}^{n} x_i - k = 0 \tag{35}$$

$$-\sum_{j=1}^{n} a_{ij}x_j + s_i + y_i = 0, \text{for } i = 1, ..., n \tag{36}$$

$$y_i - M(1 - x_i) \leq 0, \text{for } i = 1, ..., n \tag{37}$$

$$h_i - Mx_i \leq 0, \text{for } i = 1, ..., n \tag{38}$$

$$-\sum_{j=1}^{n} d_{ij}x_j + h_i \leq 0, \text{for } i = 1, ..., n \tag{39}$$

$$\sum_{i=1}^{n} h_i \geq T_\alpha k(k-1) \tag{40}$$

where $x_i \in \{0, 1\}$ and $s_i, y_i, h_i \geq 0$, for $i, j = 1, ..., n$.

Applying CPLEX 7.0, this problem can be easily solved with $n = 30$. This formulation is very computationally efficient and is used to solve this quadratically constrained quadratic zero-one problem iteratively for the selection after every subsequent seizure. In the future, it may be useful for diagnostic purposes to implant more electrodes. Although this will increase n, this formulation is still applicable because it is computationally efficient. Note that, in the future, more seizure characteristics may be discovered. This would require additional quadratic and linear constraints, problem formulation technique is still applicable for solving MQIP problems.

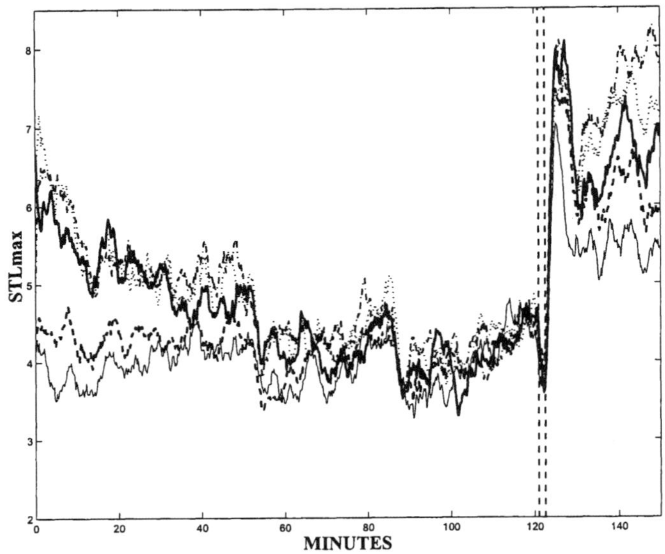

Figure 1.12. Smoothed STL_{max} profiles of 5 optimal electrode sites over 150 minutes including a seizure. The preictal period shows gradual convergence of the STL_{max} values calculated for these critical electrode sites. During the seizure, STL_{max} values are completely entrained. Postictally, the values are disentrained indicating resetting which reverses the preictal entrainment.

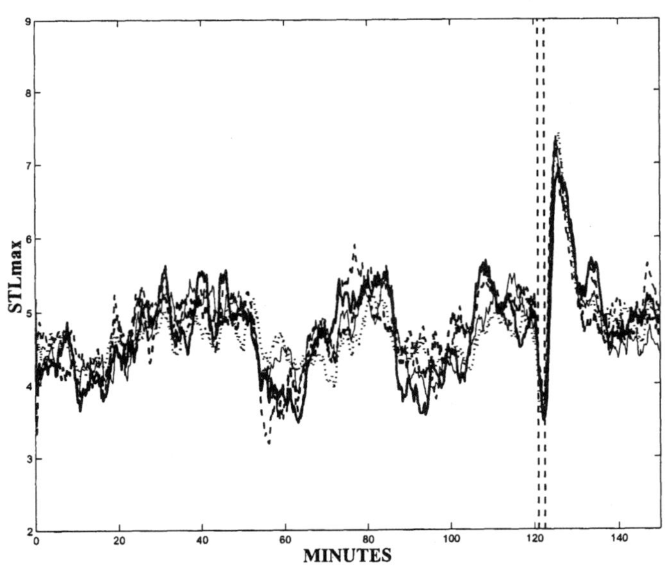

Figure 1.13. Smoothed STL_{max} profiles of 5 non-optimal electrode sites over 150 minutes including a seizure. Postictal resetting is not observed for these sites.

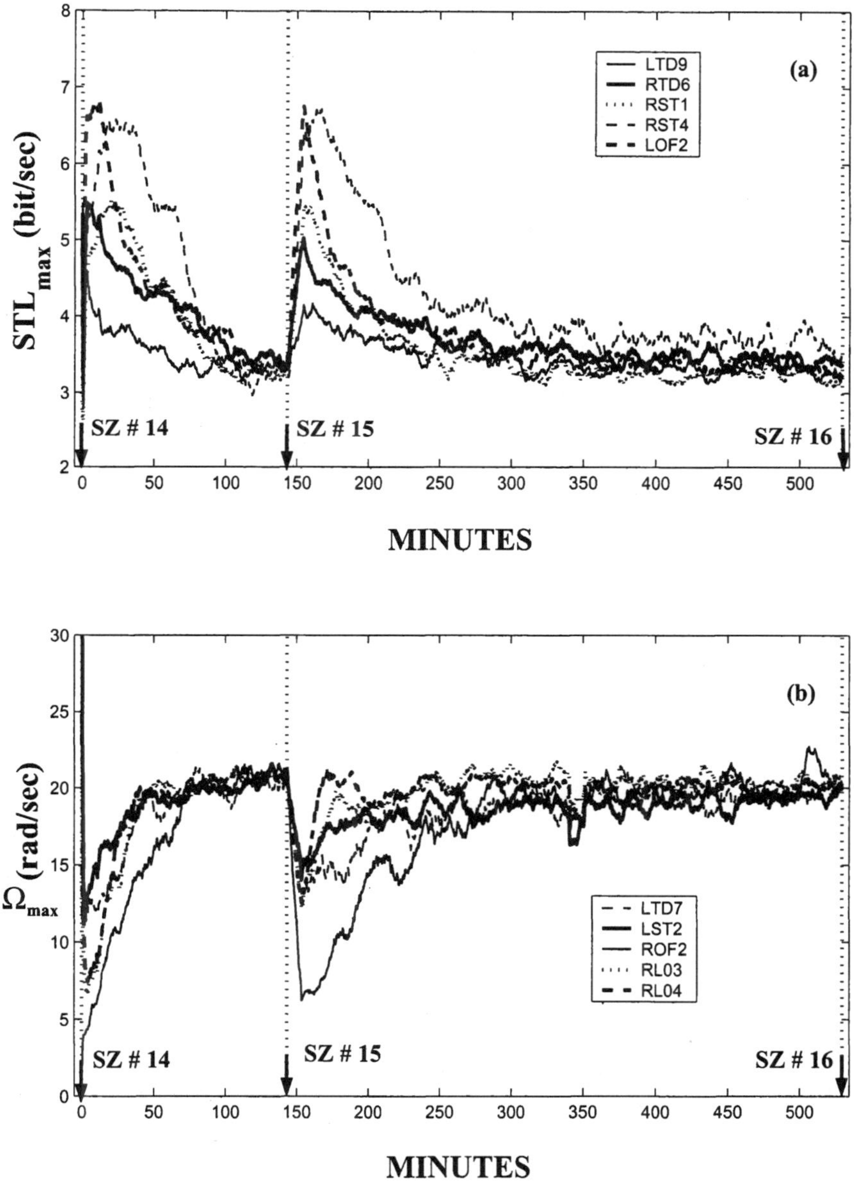

Figure 1.14. (a) Smoothed STL_{max} profiles of the 5 optimally selected electrodes over time (including seizures 14, 15, and 16). The optimal electrodes were selected 10 minutes before seizure 15. (b) Smoothed $\bar{\Omega}_{max}$ profiles from the same EEG data as in (a).

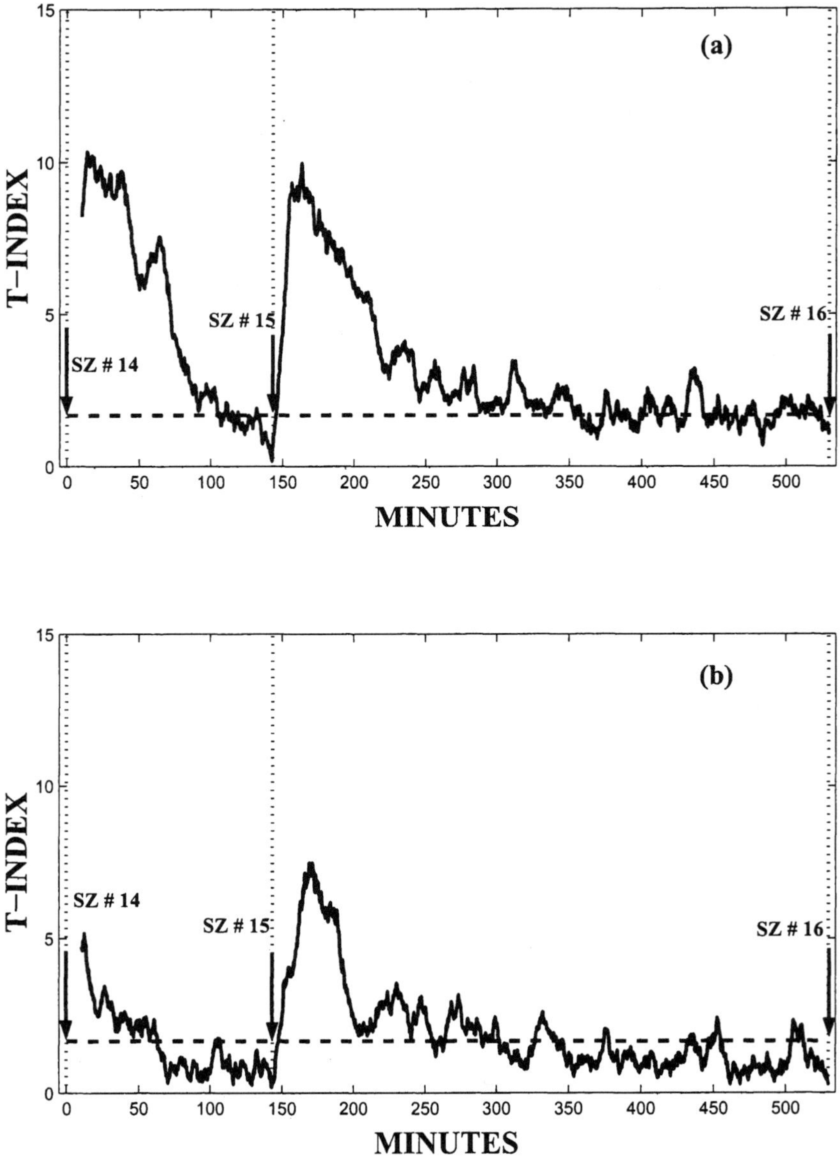

Figure 1.15. (a) Average T-index curve over time from the STL_{max} profiles in Figure 1.14(a). (b) Average T-index curve over time from the $\bar{\Omega}_{max}$ profiles in Figure 1.14(b).

For illustration purposes, the smoothed (10-minute moving average) STL_{max} and Ω_{max} profiles of the five optimally selected electrodes of seizure 14 and 15, are shown in Figures 1.14. The optimal electrodes were selected in a 10 minute interval prior to the second seizure of each set. For each set of seizures, STL_{max} and Ω_{max} profiles clearly converge (entrain) before the second seizure and either both or one of them diverge (disentrain) in this seizure's postictal period. The average T-index curves that quantify this preictal entrainment and postictal disentrainment among the selected electrodes for each of the corresponding 3 sets of seizures are respectively shown in Figures 1.15. The second and third sets of seizures were included herein to show that STL_{max} and Ω_{max} measures are not identical in the detection of the entrainment and disentrainment transition across epileptic seizures in the same patient.

Figures 1.16 and 1.17 illustrate the application of the optimization techniques to the detection of the preictal transition preceding seizure 10 in patient 1 from STL_{max} profiles. The profiles from 5 electrode sites (k=5), selected with the optimization program applied during the 10 minute interval immediately preceding the onset of seizure 9 are shown. Figures 1.18 and 1.19 illustrate the detection of the preictal transition preceding seizure 14 in patient 1 from $\bar{\Omega}_{max}$ profiles. The profiles from 5 electrode sites (k=5), selected with the optimization program applied during the 10 minute interval immediately preceding the onset of seizure 13 are shown. In Figure 1.16, the estimated values forward in time from the 5 selected sites are shown for the entire interval between seizures 9 and 10. The average T-index over all possible T_{ij} indices among the optimal 5 sites is plotted over time in Figure 1.17. In Figure 1.18, the estimated values forward in time from the 5 selected sites are shown for the entire interval between seizures 13 and 14. The average T-index over all possible T_{ij} indices among the optimal 5 sites is plotted over time in Figure 1.19.

Several points are noteworthy. First, the use of optimal sites for the estimation of and average T-index profiles helps to detect the preictal transition (about 1 to 0.5 hour before the seizure onset). Second, the detection of the preictal transition is more robust than the previous illustration when only 2 sites were used (compare Figures 1.9 and 1.10), in the sense that false warnings preceding the seizure have been eliminated. The need for optimization in selecting cortical site that are likely to show the preictal transition prior to an impending seizure is clear when we compare the results derived with optimization (e.g. in Figures 1.18 and 1.19) to sites selected without the use of optimization (e.g. in Figures 1.20 and 1.21). In Figure 1.20, the profiles from 5 randomly selected electrode sites (k=5) from the same 10 minute interval prior to seizure 13 are estimated for the same interval between seizures 13 and 14. The corresponding average T-index profile is shown in Figure 1.21. It is clear that the preictal transition of seizure 14 cannot be reliably detected by analyzing the EEG from sites that were not selected using the optimization program.

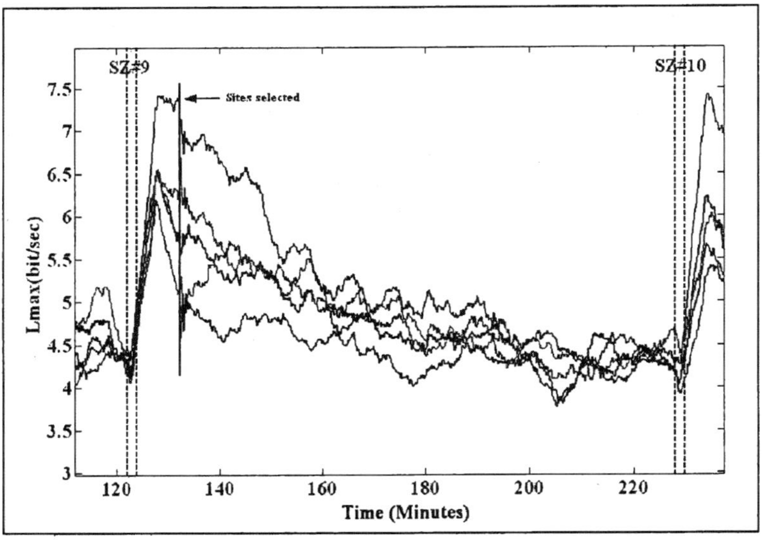

Figure 1.16. Convergence of 5 STL_{max} profiles from critical cortical sites over 2 hours between seizures 9 and 10 (patient 1). The ictal periods of the two seizures are denoted by vertical lines.

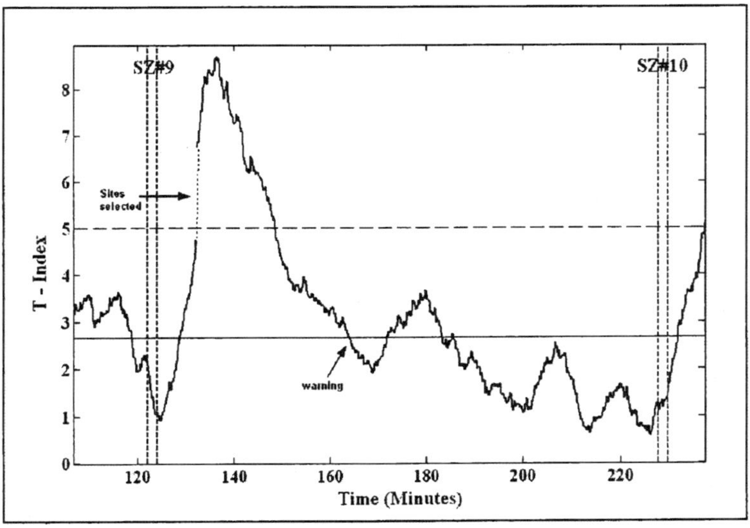

Figure 1.17. The T-index profile among 5 critical cortical sites whose STL_{max} profiles are depicted in Figure 1.16. The cortical sites are dynamically entrained approximately 60 minutes prior to seizure's 10 onset. The statistical thresholds are represented by the two horizontal lines.

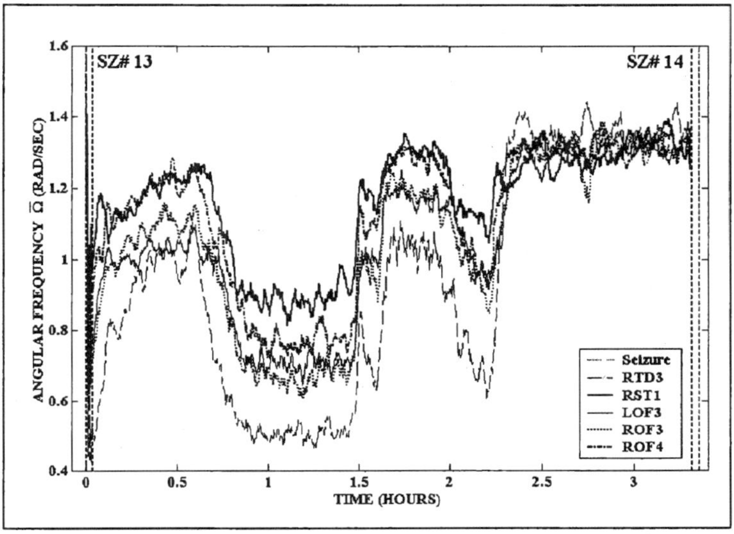

Figure 1.18. Angular frequency $\bar{\Omega}_{max}$ profiles between seizures 13 and 14 (patient 1) of 5 electrode sites selected by the optimization program during the 10 minute interval prior to the onset of seizure 13.

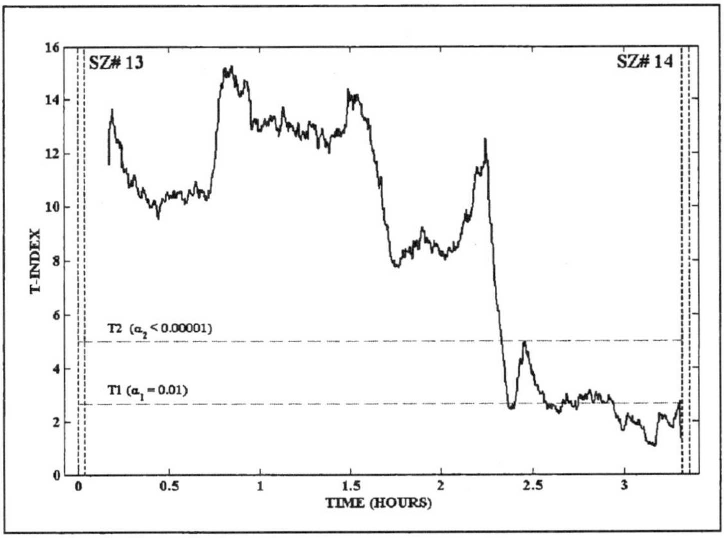

Figure 1.19. The average T-index profile of the 5 optimal electrode sites whose $\bar{\Omega}_{max}$ profiles are depicted in Figure 1.18. The 5 sites become and remain dynamically entrained approximately 0.5 hour prior to the onset of seizure 14.

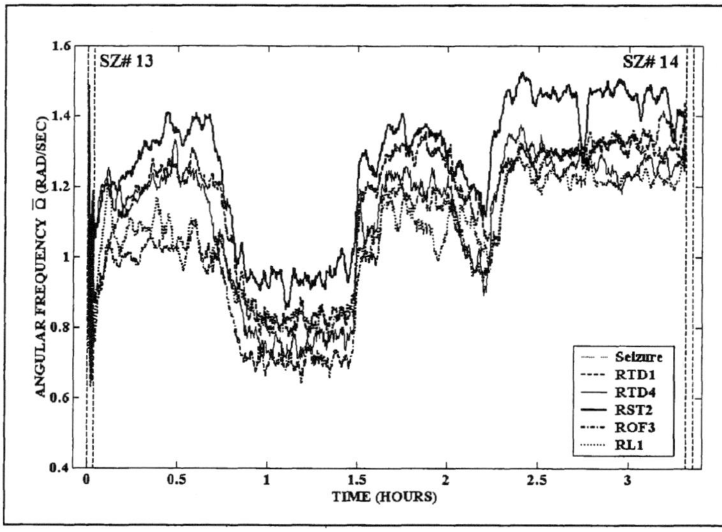

Figure 1.20. Angular frequency $\bar{\Omega}_{max}$ profiles between seizures 13 and 14 (patient 1) of 5 non-optimally selected electrode sites.

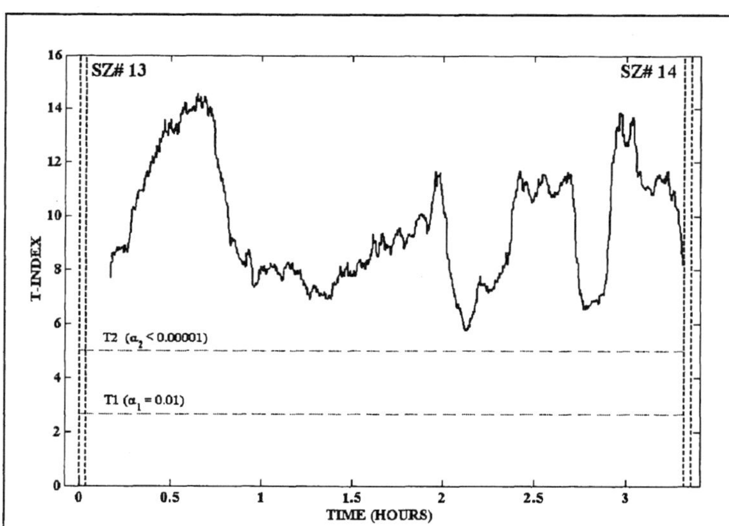

Figure 1.21. The average T-index profile of the 5 non-optimal electrode sites whose $\bar{\Omega}_{max}$ profiles are depicted in Figure 1.20. The 5 sites do not become dynamically entrained between seizures 13 and 14.

4.4. Results

Having defined that electrode sites, which participate in preictal transition, must be entrained prior to seizures, we hypothesize that "the electrode sites that are most entrained during the current seizure and disentrained after the seizure onset should be more likely to be entrained prior to the next seizure than the other entrained sites." Critical electrode sites are those sites, which are most entrained prior to seizures and disentrained after the seizure onset. As a result, it is possible to predict a seizure if one can identify critical electrode sites in advance. To test this hypothesis, we designed an experiment, which compares the probability of detecting preictal transition from any entrained cortical sites with the probability of detecting preictal transition from the critical cortical sites. In this experiment, testing on 3 patients with 20 seizures, we randomly selected 5,000 groups of entrained sites, and employ the computational approach, in the previous section, to solve multi-quadratic problem (select the critical sites).

The results show that the probability of detecting preictal transition from the critical cortical sites is approximately 83%. When we compare this probability with the probability of detecting preictal transition from any entrained sites, we obtain P-value < 0.07, which is significant and validates our hypothesis. The Histogram of probability of detecting preictal transition from randomly selected entrained cortical sites compared with the probability of detecting preictal transition from the critical cortical sites is illustrated in Figure 1.22.

5. Conclusions and Discussion

In this paper, we are interested in multi-quadratic 0-1 programming problem, which is one of the most practical optimization problems. We proposed a new computational approach to solve the multi-quadratic 0-1 programming problem. In this approach, we have developed a novel linearization technique based on Karush–Kuhn Tucker (KKT) optimality conditions. It is well-known that the KKT optimality conditions guarantee the global optimality only in the convex case. Although the developed technique seems to be heuristic in nature, we have proven that this novel technique can guarantee the global optimality with the positivity assumption of elements in the quadratic matrices. It is worth noting that because of the properties of T-index matrices, the positivity assumption always holds.

While this developed technique solves the multi-quadratic 0-1 programming problems with global optimality, it linearizes the problem with the same number of 0-1 variables (n) and additional $O(n)$ number of continuous variables. On the other hand, the conventional linearization techniques found in the literature linearize the problem with additional $O(n^2)$ number of 0-1 variables. This makes the problem become much larger and harder to solve. The comparison of computational times between the conventional linearization approach (found

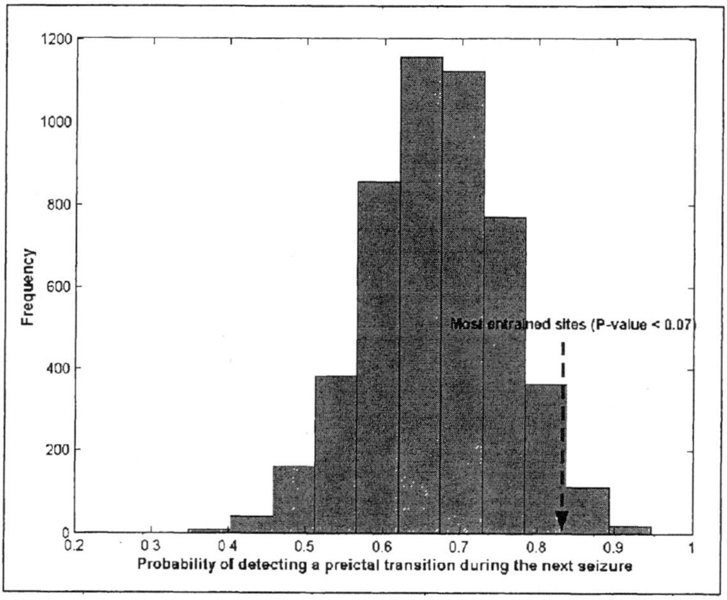

Figure 1.22. Histogram of probability of detecting preictal transition of randomly selected entrained electrode sites 5,000 times compared with the most entrained electrode sites.

in the literature) and the KKT conditions linearization approach has shown that this developed technique enormously outperforms the conventional approach; that is, the new technique solves problems a lot faster than the conventional one and consume considerably much less computational resources.

The results of this study in epilepsy confirm our hypothesis that the set of most converged cortical sites during the current seizure and reset after seizure onset is more likely to be converged again during the next seizure than other converged cortical sites. These results indicate that it may be possible to develop automated seizure warning devices for diagnostic and therapeutic purposes. Thus, it is possible to predict an impending seizure based on optimization and nonlinear dynamics of multichannel intracranial EEG recordings. Prediction is possible because, for the vast majority of seizures, the spatiotemporal dynamical features of the preictal transition are sufficiently similar to that of the preceding seizure. This similarity makes it possible to identify electrode sites that will participate in the next preictal transition, by solving multi-quadratic 0–1 problem. Although evidence for the characteristic preictal transition utilized by the seizure prediction algorithm employed in this study was first reported by our group in 1991 [11], further studies were required before a practical seizure prediction algorithm was feasible. Development of a seizure prediction algorithm was complicated because the cortical sites participating in the preictal transition varied from seizure to seizure. This problem was overcome by the use of our proposed approaches to solve multi-quadratic 0-1 problem. Because the algorithm selects candidate electrode sites and by analyzing continuous EEG recordings of several days of duration, the computational approach to solve the optimization problem has to be very efficient. At present, we can efficiently solve multi-quadratic 0-1 problem. However, future technology may allow physicians to implant thousands of electrode sites, $n > 1000$, in the brain. This procedure will extract more information and allow us to have a more understanding about the brain. Therefore; to solve this optimization problem with $n > 1000$, we may need computationally fast heuristic approaches in the future.

References

[1] G.G. Athanasiou, C.P. Bachas, and W.F. Wolf. Invariant geometry of spin-glass states. *Physical Review B*, 35:1965–1968, 1987.

[2] F. Barahona. On the computational complexity of spin glass models. *J. Phys. A: Math. Gen.*, 15:3241–3253, 1982.

[3] F. Barahona. On the exact ground states of three-dimensional ising spin glasses. *J. Phys. A: Math. Gen.*, 15:L611–L615, 1982.

[4] L. Diambra, J.C. Bastos de Figueiredo, and C.P. Malta. Epileptic activity recognition in eeg recording. *Physica A*, 273:495–505, 1999.

[5] C.E. Elger and K. Lehnertz. Seizure prediction by non-linear time series analysis of brain electrical activity. *European Journal of Neuroscience*, 10:786–789, 1998.

[6] W.G. Frank, T. Lookman, M.A. Nerenberg, C. Essex, J. Lemieux, and W. Blume. Chaotic time series analyses of epileptic seizures. *Physica D*, 46:427–438, 1990.

[7] R. Horst, P.M. Pardalos, and N.V. Thoai. *Introduction to global optimization*. Kluwer Academic Publishers, 1995.

[8] L.D. Iasemidis. *On the dynamics of the human brain in temporal lobe epilepsy*. PhD thesis, University of Michigan, Ann Arbor, 1991.

[9] L.D. Iasemidis, P.M. Pardalos, J.C. Sackellares, and D.-S. Shiau. Quadratic binary programming and dynamical system approach to determine the predictability of epileptic seizures. *Journal of Combinatorial Optimization*, 5:9–26, 2001.

[10] L.D. Iasemidis, J.C. Principe, and J.C. Sackellares. Measurement and quantification of spatiotemporal dynamics of human epileptic seizures. In M. Akay, editor, *Nonlinear biomedical signal processing*, pages 294–318. Wiley–IEEE Press, vol. II, 2000.

[11] L.D. Iasemidis and J.C. Sackellares. The evolution with time of the spatial distribution of the largest lyapunov exponent on the human epileptic

cortex. In D.W. Duke and W.S. Pritchard, editors, *Measuring Chaos in the Human Brain*, pages 49–82. World Scientific, 1991.

[12] L.D. Iasemidis, D.-S. Shiau, P.M. Pardalos, and J.C. Sackellares. Phase entrainment and predictability of epileptic seizures. In P.M. Pardalos and J.C. Principe, editors, *Biocomputing*, pages 59–84. Kluwer Academic Publishers, 2001.

[13] L.D. Iasemidis, D.-S. Shiau, J.C. Sackellares, and P.M. Pardalos. Transition to epileptic seizures: Optimization. In D.Z. Du, P.M. Pardalos, and J. Wang, editors, *DIMACS series in Discrete Mathematics and Theoretical Computer Science*, pages 55–74. American Mathematical Society, 1999.

[14] L.D. Iasemidis, D.-S. Shiau, J.C. Sackellares, P.M. Pardalos, and A. Prasad. Dynamical resetting of the human brain at epileptic seizures: application of nonlinear dynamics and global optimization tecniques. *IEEE Transactions on Biomedical Engineering*, To appear, 2003.

[15] L.D. Iasemidis, H.P. Zaveri, J.C. Sackellares, and W.J. Williams. Phase space topography of the electrocorticogram and the lyapunov exponent in partial seizures. *Brain Topography*, 2:187–201, 1990.

[16] K. Lehnertz and C.E. Elger. Can epileptic seizures be predicted? evidence from nonlinear time series analysis of brain electrical activity. *Phys. Rev. Lett.*, 80:5019–5022, 1998.

[17] B. Litt, R. Esteller, J. Echauz, D.A. Maryann, R. Shor, T. Henry, P. Pennell, C. Epstein, R. Bakay, M. Dichter, and G. Vachtservanos. Epileptic seizures may begin hours in advance of clinical onset: A report of five patients. *Neuron*, 30:51–64, 2001.

[18] M. Mezard, G. Parisi, and M.A. Virasoro. *Spin glass theory and beyond.* World Scientific, 1987.

[19] P.M. Pardalos and G. Rodgers. Parallel branch and bound algorithms for unconstrained quadratic zero-one programming. In R. Sharda et al., editor, *Impact of recent computer advances on operations research*. North-Holland, 1989.

[20] P.M. Pardalos and G. Rodgers. Computational aspects of a branch and bound algorithm for quadratic zero-one programming. *Computing*, 45:131–144, 1990.

[21] M. Le Van Quyen, J. Martinerie, M. Baulac, and F. Varela. Anticipating epileptic seizures in real time by non-linear analysis of similarity between eeg recordings. *NeuroReport*, 10:2149–2155, 1999.

[22] J.C. Sackellares, L.D. Iasemidis, R.L. Gilmore, and S.N. Roper. Epileptic seizures as neural resetting mechanisms. *Epilepsia*, 38(S3):189, 1997.

[23] J.C. Sackellares, L.D. Iasemidis, R.L. Gilmore, and S.N. Roper. Epilepsy - when chaos fails. In K. Lehnertz, J. Arnhold, P. Grassberger, and C.E. Elger, editors, *Chaos in the brain?* World Scientific, 2002.

[24] J.C. Sackellares, L.D. Iasemidis, and D.-S. Shiau. Detection of the preictal transition in scalp eeg. *Epilepsia*, 40:176, 1999.

[25] D.-S. Shiau, Q. Luo, S.L. Gilmore, S.N. Roper, P.M. Pardalos, J.C. Sackellares, and L.D. Iasemidis. Epileptic seizures resetting revisited. *Epilepsia*, 41(S7):208–209, 2000.

[26] A. Wolf, J.B. Swift, H.L. Swinney, and J.A. Vastano. Determining lyapunov exponents from a time series. *Physica D*, 16:285–317, 1985.

Chapter 2

NONLINEAR NEURODYNAMICAL FEATURES IN AN ANIMAL MODEL OF GENERALIZED EPILEPSY*

P.R. Carney
Department of Pediatrics, Neurology, and Neuroscience, University of Florida
carnepr@peds.ufl.edu

D.-S. Shiau
Department of Neuroscience, University of Florida; VA Medical Center, Gainesville, FL
shiau@epilepsy.health.ufl.edu

P.M. Pardalos
Departments of Industrial and Systems Engineering, Computer Science, and Biomedical Engineering, University of Florida
pardalos@cao.ise.ufl.edu

L.D. Iasemidis
Department of Biomedical Engineering, Arizona State University
Leon.Iasemidis@asu.edu

W. Chaovalitwongse
Department of Industrial and Systems Engineering, University of Florida; VA Medical Center, Gainesville, FL
arty@epilepsy.health.ufl.edu

*This work was supported by the Department of Veterans Affairs and grants from NIH/NINDS (RO1-NS-31451), NIH/NIBIB (8R01EB002089-03), University of Florida Division of Sponsored Research, and the Children's Miracle Network.

J.C. Sackellares
Departments of Neuroscience, Neurology, Pediatrics, and Biomedical Engineering, University of Florida; VA Medical Center, Gainesville, FL
sackellares@mbi.ufl.edu

Abstract Epileptic seizures result from the intermittent spatial and temporal summation of abnormally discharging neurons [3, 7, 20, 24]. Complex dynamical interactions between brain regions act to recruit and entrain neurons by loss of inhibition and synchronization. Long-term continuous 4-channel intracranial electroencephalographic (EEG) recordings were obtained from a genetically engineered model of generalized epilepsy ($n = 3$) and littermate controls ($n = 3$) in order to perform nonlinear dynamical analyses of intracranial brain electrical activity. Signal processing techniques included reconstruction of the EEG signal as trajectories in a phase space, applying nonlinear indicators on the trajectories (i.e., short-term maximum Lyapunov exponent), and statistical index (T-index) for quantifying interactions among distant brain sites. Analysis of interictal (seizure-free) and seizure prone periods in 2 to 3 week old H218 epileptic knockout mice revealed that (1) the brain electrical activity is of higher order during the seizure-prone period, and (2) the interaction among brain sites is more active during the seizure-prone period. In addition, dynamical analyses do not show significant difference between interictal periods in epileptic mice and littermate controls. These results suggest that the development of seizures in an animal model of generalized epilepsy is determined in part by non-stochastic neural processes. Results further suggest that it may be possible to identify the occurrence of seizures in advance through dynamical analytic distinction of interictal and seizure-prone periods.

Keywords: Epilepsy, nonlinear dynamics, Lyapunov exponents, T-index, animal models

1. Introduction

Seizure onset results from the spatial and temporal summation of abnormally discharging neurons, which act to recruit and entrain other neurons by loss of inhibition and synchronization [3,7,20,24]. Thus, preictal changes should be detectable during the period of neuronal recruitment and entrainment. Traditional linear signal analyses, such as frequency coherence [9], focal spike density counts [10], or spectral analyses are not reliable indicators for seizure prediction. Nonlinear indicators have been shown to undergo predictable changes in advance of seizure onset [8, 11–15, 18, 19]. These studies suggest that seizures represent the spontaneous formation of self-organizing spatial and temporal patterns of brain activity. The interictal to ictal state transition denotes a gradual phase transition from a complex to a less complex (more ordered) state during the preictal phase. This preictal transition was detected in human partial epilepsy using techniques developed for the study of complex non-linear systems [1].

We undertook the present study to determine whether nonlinear analyses of continuous EEG could distinguish between interictal and preictal states in an experimental animal model of generalized epilepsy in which the entire protein coding region of the single copy mouse H218 gene was deleted through homologous recombination [17]. Southern blot analysis with 3' and 5' probes, as well as PCR analysis confirmed the appropriate location of the mutation in both ES cells and mice following germ line transmission. The appearance and behavior of the newborn $H218^{-/-}$ mice were indistinguishable from that of their $H218^{+/+}$ littermates. The loss of H218 had no effect on the weight or length of the mice throughout development. Necropsy of adult $H218^{+/+}$ and $H218^{-/-}$ mice did not reveal any differences in general anatomy of the position, size, or appearance of major organs. Likewise, examination of cresyl violet-stained sections in embryonic, newborn, and postnatal day (P) 18-35 mice brains did not detect any genotype dependent differences. In particular, the $H218^{-/-}$ mice did not display any gross abnormalities reflecting potential defects in neuronal proliferation, migration, differentiation or survival. Recurrent and unprovoked spike-and-wave discharges (SWDs) occurred beginning at the end of the second postnatal week [17]. SWDs were often accompanied by stereotypical ictal behavioral changes, including frozen staring, vibrissae twitching, facial myoclonus, inability to move, and wild running fits [17]. Electroencephalographic (EEG) seizures, ictal behavioral manifestations, timing of seizure onset and termination, varied little between siblings, litters, and generations. These intrinsic characteristics of H218 were deemed as highly suitable in order to conduct our dynamical nonlinear analyses of cortical neuronal activity.

Spatiotemporal information of brain activity was obtained from multi-electrode, continuous, 24-hour recordings in P18-25 H218 mice and age-matched littermate controls. Our objective was to follow the transition toward epileptic seizures in H218 mice by reconstructing EEG recordings as trajectories in a phase space, applying non-linear indicators on such trajectories, and utilizing statistical index to quantify the interactions among different brain sites. An important concept when studying the dynamics of a system using nonlinear analyses is the reconstruction of the phase or state space. The phase space of a dynamical system is a mathematical space with orthogonal coordinate directions representing each of the variables needed to specify the instantaneous state of the system [4]. The phase space reconstruction method uses data in order to construct vectors by iterations of a time delay [1]. Here, we use 7 sequential voltages, with 20 milliseconds delay, in order to generate a point in 7-dimensional phase space. The process is repeated to generate all the possible points within in each 10.24-second epoch at a sampling frequency of 200 Hz. These points form an attractor. The Lyapunov exponent is a numerical indicator that describes the average rate at which the trajectories of adjacent states in the phase space diverge or converge over time [1].

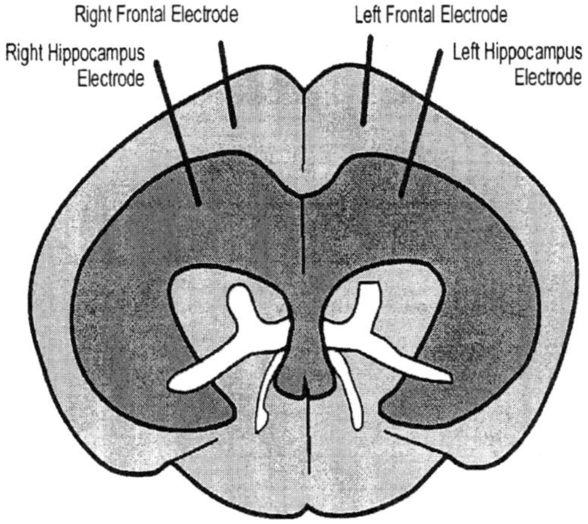

Figure 2.1. Depth electrode placement diagram. Microelectrodes are placed in the right and left frontal cortex and right and left hippocampus in order to record continuous brain electrical activity.

Previous studies in epileptic patients revealed that the preictal state is characterized by a gradual convergence of the short-term maximum Lyapunov exponent (STL_{max}) values recorded from critical electrode sites to a common value [19]. This convergence has been defined as dynamical entrainment [19]. The degree of significance of the dynamical entrainment can be quantified by a statistical T-index from the standard pair-T test. In the present study, we attempt to utilize nonlinear dynamical STLmax measure and statistical T-index to investigate the neurodynamical difference between interictal and preictal states in epileptic mouse model with generalized seizures.

2. Materials and Methods

2.1. The H218 model

Untimed pregnant $H218^{-/-}$ mice were housed at the animal facility at the University of Florida (Gainesville, FL). The sucking mice were weaned at P17 and then housed in groups of three animals of the same gender, until the day of surgery for electrode implantation. All animals were maintained in a controlled environment at 12-hr light, 12-hr dark cycle with lights on at 0600 hr. They were given ad lib access to food and water.

2.2. Surgery and video-electrocorticography

Chronic intracranial microelectrode implantation in $H218$ ($n = 3$) and wild type ($n = 3$) littermates allowed for continuous video electroencephalographic recordings along with simultaneous video-monitoring [17]. $P17$ $H218$ mice (average body weight for males and females was $19 - 25$ g) were anesthetized with avertin (1.25% tribromoethanol/ amyl alcohol solution, 0.02 ml/gm, i.p.) which lasted for $0.45 - 1$ hr, and placed in a Kopf stereotactic frame. The scalp was split and all soft tissue loosened from the dorsum of the skull. Microelectrodes (0.125 mm diameter, platinum; Plastics One, Inc., Roanoke, VA) were chronically implanted 1 mm deep into the right and left dorsal hippocampus and right and left frontal cortex (See Fig. 2.1). The points of bregma, lambda, and the auditory meatus were used for reference. A correction factor for the stereotaxic coordinates for mouse brain was calculated according to the regression equation $F' = Fa - 0.66(bl - 3.8)$, where F' is the predicted frontal coordinate of a particular neural point, bl is bregma, and Fa is the frontal coordinate of that point given by a mouse brain atlas [21]. Additional reference and ground electrodes were also used. All electrodes, intracranial, reference and ground, were connected to a molded plastic pedestal (Plastics One, Inc., Roanoke, VA) which was secured to the skull by cranioplastic dental cement. $H218^{-/-}$ and wild-type controls were operated on, recorded and evaluated in parallel. Animals were allowed to recover from anesthesia for 24-hr before returning to a standard mouse cage.

On the day of recordings ($P18 - 25$), the mice were freely moving in individual, warm, Plexiglas chambers (Dragonfly, WA). All animals were maintained in a controlled environment at 12-hr light, 12-hr dark cycle with lights on at 0600 hr in order to minimize circadian variations. They were given ad lib access to food and water. A 20-min adaptation period before electrocortigraphy recordings minimized movement artifact. Four-channel monopolar EEG recordings were made on a 32-channel digital recorder (BSMI/Nicolet 4000, Madison, WI, U.S.A.). Referential, unilateral left and right, bipolar and linked monopolar recordings from ipsilateral regions were compared to determine the phase relation, amplitudes, and source of the spike and wave discharges [17].

EEGs were recorded on-line (digitally) onto high fidelity videotape. Prior to storage on the magnetic medium, signals were sampled at 200 Hz, using an analog to digital (A/D) converter with 12- bits quantitation, and amplifiers with input range of -2.5 to +2.5 mV and frequency range of 0.05 Hz to 70 Hz. Subsequently (off-line), the data from the magnetic tapes was transferred to the hard disk (maximum capacity 9 Gigabyte), and finally to optical disks (capacity 1.5 Gigabyte). Computer data files were created that included rodent identification, time and date of each seizure, the classification of each seizure and the original EEG data. Mice were able to move freely around a cage containing food and water during the EEG recording sessions. During recordings, animals were observed directly by one or two investigators and suspected behavioral seizure activity was recorded in a separate log. EEG recordings were compared with concomitant animal behavior by using a split screen monitor display of the video-EEG record. Video records were reviewed daily for evidence of SWDs and behavioral seizures.

All animal procedures were reviewed and approved by the University of Florida Institutional Animal Care and Use Committee.

2.3. Seizures in H218 mice

$P18-25$, $H218$ mice manifested behavioral seizures that consisted of frozen staring, vibrissae twitching, facial myoclonus, inability to move, and less frequently, wild-running/bouncing fits. In the present study, 168 behavioral seizures were time-locked with SWDs in $P18-25$, $H218$ mice ($n = 3$). The most severe seizures were succeeded by wild-running, bouncing fits followed by behavioral and EEG depression. Eight running/bouncing seizures were identified in $H218$ mice between $P18-25$ (4.2 running, bouncing fits/$H218$ mouse).

EEGs of H218 mice had relatively normal backgrounds during maximal alertness. During periods of behavioral quiescence generalized sharp waves were often seen. Obvious seizure behavior (i.e., frozen staring, vibrissae twitching, facial myoclonus, inability to move, and less frequently, wild-running/bouncing fits) was always accompanied by SWDs. In some instances, SWDs were observed in the absence of overt ictal behavior. Seizures often occurred in clusters over a 0.5-hr period. Individual seizure duration varied from 7 to 20-sec. Average seizure duration was 12-sec. A circadian pattern to seizure occurrence was not observed in this limited cohort.

Electrographic traces of seizure episodes often began with the synchronous buildup of 4-5 Hz, 30-60 μV amplitude SWDs. During the behavioral ictus, bilateral 5-6 Hz, 60-100 μV SWDs were noted. An example EEG trace recorded during a motor seizure from one $H218$ animal is shown in 2.2a,b,c. During behavioral sleep, individual spikes and runs of 2 to 3 second bilateral spiking were seen. SWDs usually disappeared on arousal. Wild-type mice did not exhibit overt seizures though occasional behavioral arrests were seen in 2 mice. No EEG abnormalities were identified during wakefulness and sleep in wild-type littermates (see Fig. 2.2d).

2.4. Short-term maximum Lyapunov exponents

We utilized an estimate of the Short-Term Maximum Lyapunov exponent (STL_{max}) as the dynamical measure of the electroencephalogram. Estimation of STL_{max} was calculated by dividing the EEG signal into non-overlapping segments of 10.24-sec each. The largest Lyapunov exponent (L_{max} or L_1) is

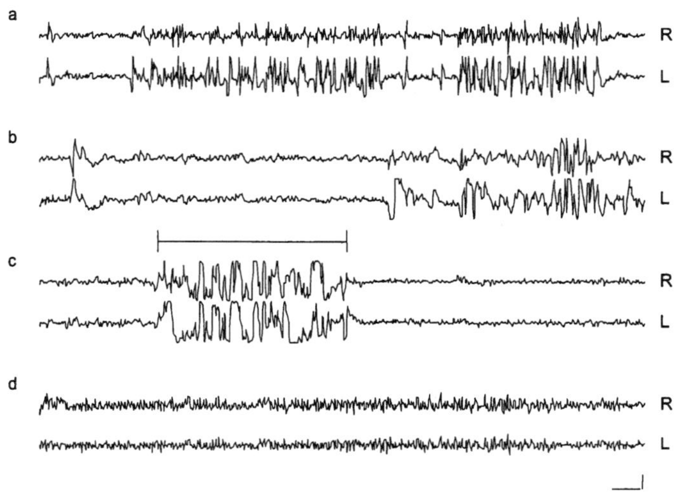

Figure 2.2. Electroencephalographic (EEG) recordings: (a) Intermittent, high amplitude, asynchronous and asymmetric, bilateral, polyspike-wave discharges in a 21 postnatal day H218$^{-/-}$ mouse during a myoclonic seizure. (b) The first 20 sec of the trace in (a) is shown at a slower time scale to better illustrate the polyspike spike/wave characteristics of the discharges. (c) Continuous, high amplitude, bilateral, polyspike-wave discharges during a wild running episode (overlying line) by a H218-/- mouse. (d) Awake H218$^{+/+}$ control mouse recording obtained using identical procedures to that of (a)-(c). Data shown are all bipolar recordings from the right (R) and left (L) frontal cortex. Similar, but less robust, results were obtained with bilateral hippocampal electrodes. Scale bar: (a), 4 sec, 30 mV; (b)-(d), 1 sec, 35 mV. (From [17]).

defined as the average of local Lyapunov exponents L_{ij} in the state space, that is:

$$L = \frac{1}{N_a} \sum_{N_a} L_{ij} \tag{1}$$

where N_a is the total number of the local Lyapunov exponents that are estimated from the evolution of adjacent points (vectors) in the state space, $X_i = X(t_i)$, $X_j = X(t_j)$, and

$$L_{ij} = \frac{1}{\Delta t} \log_2 \frac{|\delta X_{i,j}(0)|}{|\delta X_{i,j}(\Delta t)|} \tag{2}$$

where Δt is the evolution time allowed for the vector difference $|\delta X_{i,j}(\Delta t)|$ to evolve to the new difference $|\delta X_{i,j}(0)|$ where

$$\begin{aligned} \delta X_{i,j}(0) &= X(t_i) - X(t_j) \\ \delta X_{i,j}(\Delta t) &= X(t_i + \Delta t) - X(t_j + \Delta t). \end{aligned}$$

If Δt is given in sec, then L_{max} will be in bits/sec.

2.5. Statistical T-index

We used the T-index (from the statistical paired-T test) to measure the degree of entrainment between critical electrode sites. The T-index of a pair of electrodes was calculated in each 5-min epoch (30 STL_{max} segments) by dividing the mean difference of STL_{max} between the two electrode sites by its standard deviation. That is, the T-index at time t between electrode sites i and j is defined as:

$$T_{i,j}(t) = \sqrt{N} \times |E\{STL_{max,i} - STL_{max,j}\}| / \sigma_{i,j}(t) \tag{3}$$

where $E\{\cdot\}$ is the sample average difference for the $STL_{max,i} - STL_{max,j}$ estimated over a moving window $w_t(\lambda)$ defined as:

$$w_t(\lambda) = \begin{cases} 1 & \text{if } \lambda \in [t - N - 1, t] \\ 0 & \text{if } \lambda \notin [t - N - 1, t], \end{cases}$$

where N is the length of the moving window. Then, $\sigma_{i,j}(t)$ is the sample standard deviation of the STL_{max} differences between electrode sites i and j within the moving window $w_t(\lambda)$.

2.6. Statistical Analyses

STL_{max} and T-index were calculated over time in the 4-channel EEG recordings (left and right frontal cortex and left and right hippocampus) from three epileptic mice and three littermate controls. One interictal period (at least 8 hours away from a seizure) and one seizure prone period (at least 3 seizures) from each epileptic mouse were analyzed. In one epileptic mouse, 6-hours of continuous EEG recording during an interictal period, and a 2.7-hour continuous EEG recording during a seizure-prone period with 7 severe seizures was analyzed. In a second epileptic mouse, 2.9-hours of continuous EEG recording during an interictal period, and a 5.4-hours of continuous EEG recording in a seizure-prone period with 7 severe seizures was analyzed. In a third epileptic mouse, 3.6-hours of continuous EEG recording during an interictal period, and a 6-hours of continuous EEG recording during a seizure-prone period with 7 severe seizures was analyzed. These data were compared with age-matched littermate controls with 6 hours of continuous EEG recordings.

The difference of STL_{max} values between interictal and preictal periods in epileptic mice was evaluated by the two-way ANOVA with replicates test. Thirty random samples of STL_{max} values in each period were used in the test. The same analysis was also performed to compare the dynamical difference between interictal periods in H218 epileptic mice and littermate controls. Further, we compared the degree of dynamical entrainment between interictal and preictal states in H218 mice by their average T-index curves. We first defined the significant entrainment (SE) a a T-index value less than a critical value from the T-distribution with significance level = 0.05. The probabilities of significant entrainment for all periods were then estimated and compared by calculating the proportions of observed SE.

3. Results

Figures 2.3 and 2.4 show the STL_{max} 5-point smoothed curves of 4 electrodes during the interictal and preictal periods in an epileptic mouse. Figure 2.5 shows the STL_{max} 5-point smoothed curves of 4 electrodes for a littermate control. The results of the test between interictal and preictal periods in three epileptic mice showed that, for all electrodes, the STL_{max} values are significantly lower during the preictal periods ($p < 0.01$). In the comparisons of the interictal periods with the littermate controls, the test showed no significant difference in frontal cortex with respect to their STL_{max} values ($p > 0.05$). However, the STL_{max} values from the hippocampus are significantly lower in controls ($p < 0.01$).

Figures 2.6 show the average T-index curves during the interictal and seizure-prone periods from the same epileptic mouse in Figures 2.3 and 2.4, and the same littermate control in Figure 2.5. From these three T-index curves, it is obvious that the degree of entrainment is larger during the seizure-prone period. The combined paired comparison over three epileptic mice showed that the probabilities of significant entrainment (SE) during the seizure prone periods are approximately 12% higher than during the interictal periods. Further, the probability of SE in littermate control is only approximately 3% lower than the one during the interictal period in epileptic mice. However, the analysis of comparisons from more mice is required in order to show the statistical significance.

4. Discussion

The present findings indicate that it is possible to anticipate a seizure vulnerable period in a genetically engineered animal model of generalized epilepsy. These results are based on the dynamical nonlinear time-series analyses of continuous brain electrical activity collected over several days in H218 mice. By implication, the results suggest that development and resolution of seizures is determined by non-stochastic processes and that it is possible to anticipate the occurrence of seizures in advance. Similar results have been reported in human temporal lobe epilepsy [8, 14, 15, 18]. The analysis of the spatial and temporal pattern dynamics of long-term intracranial EEGs, recorded for clinical purposes in patients with medically intractable temporal lobe epilepsy, has demonstrated that seizures are preceded by dynamical changes that are detectable within 30 minutes to 1 hour before the seizure onset [11, 12, 14]. These investigations utilize the STL_{max} to quantify the rate of convergence or divergence of

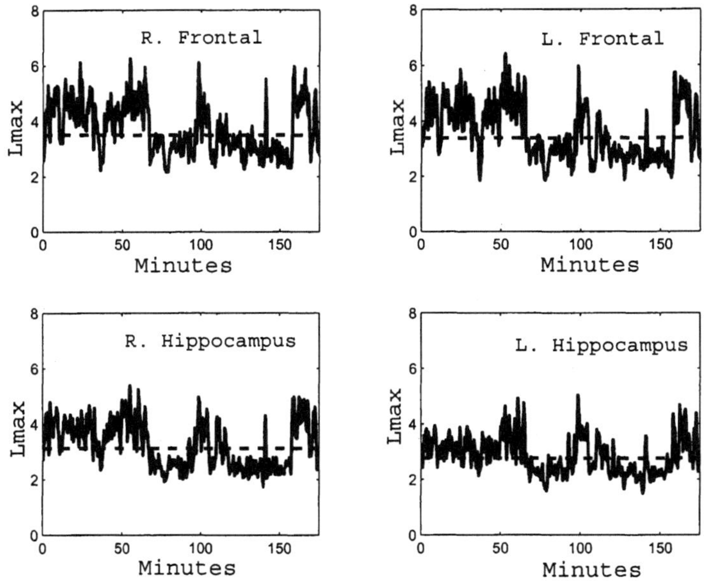

Figure 2.3. Five-point smoothed STL_{max} profiles over 2.5 hours for right frontal, left frontal, right hippocampus, and left hippocampus in a H218$^{-/-}$ mouse during the interictal (between seizure) period. The dashed horizontal lines indicate the mean STL_{max} values over the period.

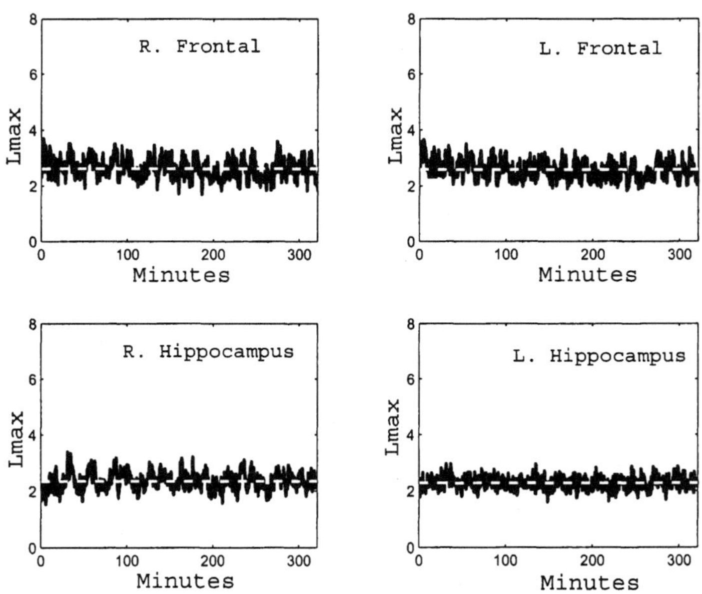

Figure 2.4. Five-point smoothed STL_{max} profiles over 5.4 hours for right frontal, left frontal, right hippocampus, and left hippocampus in a H218$^{-/-}$ mouse during the seizure-prone period. The dashed horizontal lines indicate the mean STL_{max} values over the period.

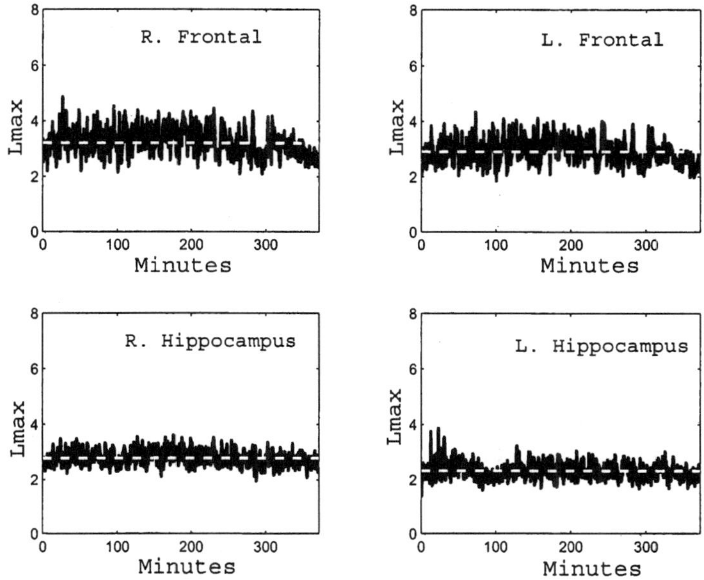

Figure 2.5. Five-point smoothed STL_{max}profiles over 6 hours for right frontal, left frontal, right hippocampus, and left hippocampus in a littermate control mouse. The dashed horizontal lines indicate the mean STL_{max}values over the period.

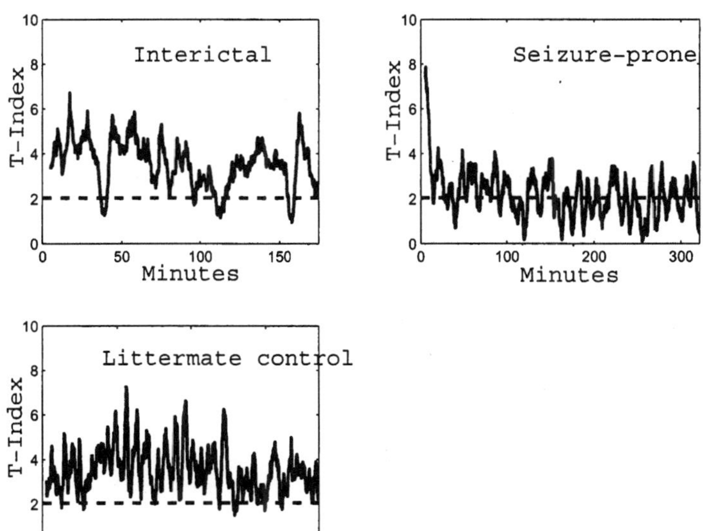

Figure 2.6. Representative example of a T-index curve among 4 electrode sites over time from interictal and seizure-prone periods in H218$^{-/-}$ and an H218$^{+/+}$ littermate control. The dashed horizontal line indicates the critical value of T-index.

neighborhood trajectories in an attractor. The preictal transition in a sample of temporal lobe seizures was characterized by a gradual convergence of STL_{max} values recorded from critical electrode sites to a common value.

What may account for the observed spatiotemporal dynamical changes that occur before seizure onset? One possibility is that seizures are preceded by physiological changes that are reflected in the dynamical characteristics of the EEG signals. The short-term Lyapunov exponent is a direct measure of the degree of order within the signal. Our findings suggest that brain dynamics are less ordered in H218 epileptic mice as compared to age-matched controls, and that the brain electrical activity becomes increasingly more ordered immediately before the seizure. For instance, STL_{max} and T-index values are lower in H218 mice as compared to age-matched controls. These results suggest that the neurodynamics of H218 epileptic mice are potentially more conducive for seizures. Seizure onset which results from the spatial and temporal summation of abnormally discharging neurons act to recruit and entrain other neurons by loss of inhibition and synchronization [3,7,20,24]. Our quantitative results support the hypothesis that the nonlinear dynamical changes are likely to represent the preictal period of neuronal recruitment and entrainment.

What are the implications of these observations? These results suggest that nonlinear quantitative analysis of broad regions of brain structures may be a more sensitive method to detect alterations in the behavior of the network before the more traditional seizure discharge is seen, and allow prediction of seizures before expression behaviorally or on traditional EEG. These results are also consistent with the suggestion that epilepsy is a disorder of large neural networks and that electrical hyperexcitability associated with seizure activity reverberates within the neural structures of the network, which operate together and inextricably to culminate in the eventual expression of a seizure by entrainment of this large neural network from any given part of the brain [22]. Interestingly, our results demonstrate that in the generalized-onset seizure model and human partial-onset seizures [11, 12, 14], seizures could be anticipated by several minutes by automated nonlinear analysis of the dynamical characteristics of intracranial EEG recordings. The similarities may reflect the existence of specific cortical and sub cortical networks in the genesis and expression of partial and generalized seizures [2, 5, 6, 16, 23], which in turn are expressed in the quantitative nonlinear values of the EEG. The present study represents a proof-of-concept preclinical effort which is a necessary initial step in the development of a seizure prediction device for humans. A seizure warning system could be incorporated into a digital signal processing chip for use in implantable devices. Such devices could be utilized to activate pharmacological or physiological therapeutic interventions designed to prevent an impending seizure. Future animal studies, employing novel experimental designs and sensitivity and specificity studies will be required to investigate the therapeutic potential for implantable seizure warning devices. The results also indicate that these methods may serve to differentiate seizure vulnerability in susceptible subjects well in advance of the onset of epilepsy.

Acknowledgments

The authors want to thank Dr. A. John MacLennan of the University of Cincinnati for providing us with an H218 mouse breeding colony that was used in this study and to thank Ms. Meadow Maze, Ms. Li Zhang, Mr. Achint Srivastava and Mr. Sandeep Nair for their superb help with reviewing the animal video-EEG tapes and assistance with the animal surgeries and recordings. We would also like to thank Ms. Lucia T. Carney-Manubens for reviewing the manuscript, and to Ms. Becky Shaffer for library assistance.

References

[1] H.D.I. Abarbanel. *Analysis of observed chaotic data.* Springer-Verlag, New York, 1996.

[2] M. Avoli and P. Gloor. Interaction of cortex and thalamus in spike and wave discharges of feline generalized penicillin epilepsy. *Experimental Neurology*, 76:196–217, 1982.

[3] T.L. Babb and W.J. Brown. Neuronal, dendritic, and vascular profiles of human temporal lobe epilepsy correlated with cellular physiology in vivo. *Advanced Neurology*, 44:946–966, 1986.

[4] G.L. Baker and J.P. Gollub. *Chaotic Dynamics: An Introduction.* Camebridge University Press, Cambridge U.K., 1990.

[5] J. Bear, N.B. Fountain, and E.W. Lothman. Responses of the superficial entorhinal cortex in vitro in slices from naïve and chronically epileptic rats. *Journal of Neurophysiology*, 76:2928–2940, 1996.

[6] E.W. Bertram. Functional anatomy of spontaneous seizures in a rat model of limbic epilepsy. *Epilepsia*, 38:95—-105, 1997.

[7] W.H. Calvin, G.A. Ojemann, and A.A. Ward. Human cortical neurons in epileptogenic foci: comparison of inter-ictal firing patterns to those of epileptic neurons in animals. *Electroencephalogram and Clinical Neurophysiology*, 34(4):337–351, 1973.

[8] F.H. Lopes da Silva and J.P. Pijn. Epilepsy: Network models of generation. In M.A. Arbib, editor, *The Handbook of Brain Theory and Neural Networks*, pages 367–369. MIT Press, 1995.

[9] R.D. Duckrow and S.S. Spencer. Regional coherence and the transfer of ictal activity during seizure onset in the medial temporal lobe. *Electroencephalogram and Clinical Neurophysiology*, 82:415—-422, 1992.

[10] J. Gotman, J. Ives, P. Gloor, A. Olivier, and L. Quesney. Changes in interictal eeg spiking and seizure occurrence in humans. *Epilepsia*, 23:432—-433, 1982.

[11] L.D. Iasemidis, P.M. Pardalos, J.C. Sackellares, and D.-S. Shiau. Quadratic binary programming and dynamical system approach to determine the predictability of epileptic seizures. *Journal of Combinatorial Optimization*, 5:9–26, 2001.

[12] L.D. Iasemidis and J.C. Sackellares. The evolution with time of the spatial distribution of the largest lyapunov exponent on the human epileptic cortex. In D.W. Duke and W.S. Pritchard, editors, *Measuring Chaos in the Human Brain*, pages 49–82. World Scientific, 1991.

[13] L.D. Iasemidis, D.-S. Shiau, J.C. Sackellares, and P.M. Pardalos. Transition to epileptic seizures: Optimization. In D.Z. Du, P.M. Pardalos, and J. Wang, editors, *DIMACS series in Discrete Mathematics and Theoretical Computer Science*, pages 55–74. American Mathematical Society, 1999.

[14] L.D. Iasemidis, H.P. Zaveri, J.C. Sackellares, and W.J. Williams. Phase space topography of the electrocorticogram and the lyapunov exponent in partial seizures. *Brain Topography*, 2:187–201, 1990.

[15] K. Lehnertz and C.E. Elger. Can epileptic seizures be predicted? evidence from nonlinear time series analysis of brain electrical activity. *Physical Review Letters*, 80:5019–5022, 1998.

[16] E.W. Lothman and R.C. Collins. Kanic acid induced limbic seizures: metabolic, behavioral, electroencephalographic and neuropathological correlates. *Brain Research*, 218:299–318, 1981.

[17] A.J. MacLennan, P.R. Carney, W.J. Zhu, A.H. Chaves, J. Garcia, J.R. Grimes, K.J. Anderson, S.N. Roper, and N. Lee. An essential role for the h218/agr16/edg-5/lp(b2) sphingosine 1-phosphate receptor in neuronal excitability. *European Journal of Neuroscience*, 14(2):203–209, 2001.

[18] J. Martinerie, C. Van Adam, and M. Le Van Quyen. Epileptic seizures can be anticipated by non-linear analysis. *Nature Medicine*, 4:1173–1176, 1998.

[19] J.C. Sackellares, L.D. Iasemidis, P.M. Pardalos, and D.-S. Shiau. Combined application of global optimization and nonlinear dynamics to detect state resetting in human epilepsy. In J.C. Principe P.M. Pardalos, editor, *Biocomputing*. Kluwer Academic Publishers, 2001.

[20] S.J. Shiff, K. Jerger, D.H. Duong, T. Chang, M.L. Spano, and W. L. Ditto. Controlling chaos in the brain. *Nature (London)*, 370:615–620, 1994.

[21] B.M. Slotnick and C.M. Leonard. A stereotaxic atlas of the albino mouse forebrain. Technical report, US Government Printing Office, 1975.

[22] S.S. Spencer. Neural networks in human epilepsy: evidence of and implications for treatment. *Epilepsia*, 43(3):219–227, 2002.

[23] L.E. White and J.L. Price. The functional anatomy of limbic status epilepticus in the rat, II: the effects of focal deactivation. *Journal of Neuroscience*, 13:4810–4830, 1993.

[24] A.R. Wyler and A.A. Ward. Epileptic neurones. In J.S. Lockard and A.A. Ward, editors, *Epilepsy: A window to brain mechanism*, pages 415–422. Raven, 1986.

Chapter 3

OPTIMIZATION TECHNIQUES FOR INDEPENDENT COMPONENT ANALYSIS WITH APPLICATIONS TO EEG DATA

Pando Georgiev

Andrzej Cichocki

Hovagim Bakardjian

Laboratory for Advanced Brain Signal Processing
Brain Science Institute
The Institute of Physical and Chemical Research (RIKEN)
Wako-shi, Saitama 351-0198
Japan
{georgiev, cia,hova}@bsp.brain.riken.go.jp

Abstract We present a survey and generalizations of some methods for ICA like maximization of kurtosis and algebraic cumulant methods for a combination of second and fourth order statistics by joint diagonalization of covariance and cumulant matrices depending on time delays. We describe an experiment with EEG data showing that the combination of second and fourth order statistics gives better results for detecting of eye movements.

Keywords: Independent Component Analysis, fixed point algorithm, cumulants methods, eye movement.

1. Introduction

The problem of Independent Component Analysis (ICA) and blind source separation (BSS) has received wide attention in various fields such as biomedical signal analysis and processing (EEG, MEG, fMRI), speech enhancement, geophysical data processing, data mining, wireless communications, image processing, etc. The literature about ICA and BSE problem is huge (see for instance [6], [12] and references therein).

The problem is formulated as follows: we can observe sensor signals (random variables) $\mathbf{x}(k) = [x_1(k), \ldots, x_n(k)]^T$ which are described as

$$\mathbf{x}(k) = \mathbf{A}\mathbf{s}(k) \quad k = 1, 2, \ldots \tag{1}$$

where $\mathbf{s}(k) = [s_1(k), \ldots, s_n(k)]^T$ is a vector of unknown source signals and $\mathbf{A}$ is $n \times n$ non-singular unknown mixing matrix.

Our objective is to estimate the source signals sequentially one-by-one or simultaneously assuming that they are statistically independent.

In this article we present a survey and generalization of some methods of ICA as maximization of kurtosis and some algebraic cumulant methods, and an experiments with EEG data confirming that combination of second and fourth order statistics gives better results in the task for extraction or estimation of components related to eye movements. More precisely, we give a mathematical explanation why the of maximization of the absolute value of the kurtosis gives independent components and present a generalization of the algorithm of Hyvärinen and Oja [13] for high order cumulants, which has high order order convergence rate. We present also a novel ICA algorithm combining second and fourth order statistics, based on the approximate joint diagonalization method (see [3]) for covariance and cumulants matrices with time delays and apply it for detection of eye movements.

2. Extraction via maximization of the absolute value of the cumulants

Maximization of nongaussianity is one of the basic ICA estimation principles (see [6], [12]). This principe is explained by the central limit theorem, according to which, sums of nongaussian random variables are closer to gaussian than the original ones. Therefore, a linear combination $y = \mathbf{w}^T\mathbf{x} = \sum_{i=1}^{n} w_i x_i$ of the observed mixture variables (which is a linear combination of the independent components as well, because of the linear mixing model) will be maximally nongaussian, if it equals one of the independent components. Below we give rigorous mathematical proof of this statement. The task how to find such a vector $\mathbf{w}$, which gives one independent component, and therefore should be one (scaled) row of the inverse of the mixing matrix $\mathbf{A}$, is the main task of the (sequential) ICA. We will describe an optimization problem for this task.

Define the function $\varphi : \mathbb{R}^n \to \mathbb{R}$ by

$$\varphi_p(\mathbf{w}) = \mathbf{cum}_p(\mathbf{w}^T\mathbf{x})$$

where $\mathbf{cum}_p$ means the self-cumulant of order p:

$$\mathbf{cum}_p(s) = \text{cumulant}(\underbrace{s, ..., s}_{p})$$

(see [16] for definition and properties of the cumulants). The following property of the cumulants is used essentially in derivation of fixed point algorithm [13] and its generalization below: if $s_i, i = 1, ..., n$ are statistically independent, then

$$\mathbf{cum}_p\left(\sum_{i=1}^{n} c_i s_i\right) = \sum_{i=1}^{n} c_i^p \mathbf{cum}_p(s_i). \tag{2}$$

Consider the maximization problems

OP(p) maximize $|\varphi_p(\mathbf{w})|$
under constraint $\|\mathbf{w}\| = 1$.

and

DP(p) maximize $|\psi_p(\mathbf{c})|$
under constraint $\|\mathbf{c}\| = 1$,

where $\psi_p(\mathbf{c}) = \mathbf{cum}_p\left(\sum_{i=1}^{n} c_i s_i\right)$.

Denoting $y = \mathbf{w}^T\mathbf{x}$, $\mathbf{c} = \mathbf{A}^T\mathbf{w}$ we have

$$y = \mathbf{c}^T\mathbf{s} = \sum_{i=1}^{n} c_i s_i$$

and

$$\varphi_p(\mathbf{w}) = \psi_p(\mathbf{A}^T\mathbf{w}). \tag{3}$$

Without loss of generality we may assume that the matrix $\mathbf{A}$ is orthogonal (assuming that we have made the well know preprocessing called "prewhitening", see [6] or [12]).

It is easy to see (using (3) and the orthogonality of $\mathbf{A}$) that the problems DP(p) and OP(p) are equivalent in sense that $\mathbf{w}_*$ is a solution of OP(p) if and only if $\mathbf{c}_* = \mathbf{A}^T\mathbf{w}_*$ is a solution of DP(p).

A very useful observation is the following: if a vector $\mathbf{c}$ contains only one nonzero component, say $c_{i_0} = \pm 1$, then the vector $\mathbf{w} = \mathbf{A}\mathbf{c}$ gives extraction (say $y(k)$) of the source with index i_0, since

$$\begin{aligned} y(k) &:= \mathbf{w}^T\mathbf{x}(k) \\ &= \mathbf{c}^T\mathbf{A}^T\mathbf{x}(k) \\ &= \mathbf{c}^T\mathbf{s}(k) = s_{i_0}(k) \quad \forall k = 1, 2, ... \end{aligned} \tag{4}$$

By the following lemma we show that the solutions $\mathbf{c}$ of DP(p) have exactly one nonzero element. So, we can obtain the vectors $\mathbf{w} = \mathbf{Ac}$ as solutions of the original problem OP(p), and by (4) we achieve extraction of one source.

One interesting property of the optimization problem OP(p) is that it has exactly n solutions (up to sign) which are orthonormal and any of them gives extraction of one source signal. The fixed point algorithm [13] finds one by one its solutions.

We note that the idea of maximizing of $\mathbf{cum}_4(\mathbf{w}^T\mathbf{x})$ in order to extract one source from a linear mixture is already considered in [8].

The next lemma gives a mathematical explanation of maximization of kurtosis method (applied by many authors) and its generalization, proposed here: maximization of the absolute value of even order cumulants.

Lemma 1 *Consider the optimization problem:*

$$\text{minimize (maximize) } f(\mathbf{v}) = \sum_{i=1}^{n} k_i v_i^p$$
$$\text{subject to } \|\mathbf{v}\| = 1,$$

where $p > 2$ is even and $\mathbf{v} = [v_1, ..., v_n]^T$. Denote

$$I^+ = \{i \in \{1, ..., n\} : k_i > 0\}$$

$$I^- = \{i \in \{1, ..., n\} : k_i < 0\}$$

and $\mathbf{e}_i = (0, ...0, 1, 0, ..., 0)$, (1 is the $i-$th place).

Then the points of local minimum are exactly the vectors $\pm\mathbf{e}_i, i \in I^-$ and the points of local maximum are exactly the vectors $\pm\mathbf{e}_j, j \in I^+$.

Proof. Applying the Lagrange multipliers theorem for a point of a local optimum $\overline{\mathbf{v}} = (\overline{v}_1, ..., \overline{v}_n)$, we write:

$$k_i p \overline{v}_i^{p-1} - 2\lambda \overline{v}_i = 0, \quad i = 1, ..., n, \tag{5}$$

where λ is a Lagrange multiplier.

Multiplying (5) by $\overline{v}_i$ and summing, we obtain:

$$p f_{opt.} = 2\lambda, \tag{6}$$

where $f_{opt.}$ means the value of f at the local optimum. From (5) and (6) we obtain

$$\overline{v}_i (k_i p \overline{v}_i^{p-2} - p f_{opt.}) = 0. \tag{7}$$

Hence $\overline{v}_i = 0$ if k_i and $f_{opt.}$ have different signs, and

$$\overline{v}_i \text{ is either } 0, \text{ or } \pm \left(\frac{f_{opt.}}{k_i}\right)^{\frac{1}{p-2}}, \tag{8}$$

which is the first order optimality condition for the initial problem.

Case 1. Assume that $k_{i_0} < 0$ for some index i_0 and $\overline{\mathbf{v}}$ is a local minimum. Then obviously $f_{loc.min.} < 0$. According to the second order sufficient optimality condition [1], a point $\mathbf{x}^0$ is a local minimum if

$$\mathbf{h}^T L''(\mathbf{x}^0)\mathbf{h} > 0 \quad \forall \mathbf{h} \in K(\mathbf{x}^0), \quad \mathbf{h} \neq 0, \tag{9}$$

where

$$K(\mathbf{x}^0) = \{\mathbf{h} : \mathbf{h}^T\mathbf{x}^0 = 0\}$$

is the tangent space to the constraint set at $\mathbf{x}^0$ and

$$L(\mathbf{x}) = \sum_{i=1}^{n} k_i x_i^p - \lambda(\|\mathbf{x}\|^2 - 1)$$

is the Lagrange function. The second order necessary condition [1] states that if $\mathbf{x}^0$ is a local minimum, then

$$\mathbf{h}^T L''(\mathbf{x}^0)\mathbf{h} \geq 0 \quad \forall \mathbf{h} \in K(\mathbf{x}^0). \tag{10}$$

In our case, by (6) and (8) we obtain

$$\begin{aligned} \mathbf{h}^T L''(\overline{\mathbf{v}})\mathbf{h} &= \sum_{i=1}^{n}(p(p-1)k_i\overline{v}_i^{p-2} - 2\lambda)h_i^2 \\ &= p f_{loc.min.}\left[(p-2)\sum_{i\in I} h_i^2 - \sum_{i\notin I} h_i^2\right], \end{aligned} \tag{11}$$

where I is the set of those indexes i, for which $\overline{v}_i$ is different from 0.

We will check the second order sufficient condition (9) for a local minimum for the points $\mathbf{e}_{i_0}^{\pm}$ (the first order optimality conditions (8) is obviously satisfied for $\mathbf{e}_{i_0}^{\pm}$). We have

$$K(\mathbf{e}_{i_0}^{\pm}) = \{\mathbf{h} : h_{i_0} = 0\},$$

therefore, for $\mathbf{h} \in K(\mathbf{e}_{i_0}^{\pm})$, $\mathbf{h} \neq 0$ we have

$$\mathbf{h}^T L''(\mathbf{e}_{i_0}^{\pm})\mathbf{h} > 0,$$

since $h_{i_0} = 0$ and $f_{loc.min.} < 0$, i.e. the second order sufficient condition (9) is satisfied, therefore $\mathbf{e}_{i_0}^{\pm}$ is a local minimum.

By (11) it follows that for any vector $\mathbf{v}'$ with at least two nonzero elements, say v'_{i_0} and v'_q, the quadratic form $\mathbf{h}^T L''(\mathbf{v}')\mathbf{h}$ can take negative values for some vectors $\mathbf{h} \in K(\mathbf{v}')$. For instance, taking the vector $\mathbf{h}'$ with i_0-th component equal to v'_q, q-th component equal to $-v'_{i_0}$ and the rest equal to zero, we see that $\mathbf{h}' \in K(\mathbf{v}')$, but $\mathbf{h}'^T L''(\mathbf{v}')\mathbf{h}' < 0$. Therefore, the second order necessary condition (10) for a local minimum is not satisfied for any vector with more than one non-zero elements.

Case 2. Assume that $k_j > 0$ for some index j and $\overline{\mathbf{v}}$ is a local maximum. We apply Case 1 to the function $-f$ and finish the proof. ■

3. A generalization of the fixed point algorithm

Consider the following algorithm:

$$\mathbf{w}(l) = \frac{\varphi_p'(\mathbf{w}(l-1))}{\|\varphi_p'(\mathbf{w}(l-1))\|}, \quad l = 1, 2, ..., \tag{12}$$

which is a generalization of the fixed point algorithm of Hyvärinen and Oja. The name is derived by the Lagrange equation for the optimization problem OP(p), since (12) tries to find a solution of it iteratively, and this solution is a fixed point of the operator defined by the right-hand side of (12).

The next theorem gives precise conditions for convergence of the fixed point algorithm of Hyvärinen and Oja and its generalization (12) (for a proof, see [9]).

Theorem 1 *Assume that s_i are statistically independent, zero mean signals and the mixing matrix $\mathbf{A}$ is orthogonal. Let $p \geq 4$ be a given even integer number, $\mathbf{cum}_p(s_i) \neq 0, i = 1, ..., n$ and let*

$$I(\mathbf{c}) = \arg \max_{1 \leq i \leq n} c_i \left| \mathbf{cum}_p(s_i) \right|^{\frac{1}{p-2}}.$$

Denote by W_0 the set of all elements $\mathbf{w} \in \mathbb{R}^n$ such that $\|\mathbf{w}\| = 1$, the set $I(\mathbf{A}^T\mathbf{w})$ contains only one element, say $i(\mathbf{w})$, and $c_{i(\mathbf{w})} \neq 0$, as $\mathbf{c} = \mathbf{A}^T\mathbf{w}$. Then

(a) The complement of W_0 has a measure zero.

(b) If $\mathbf{w}(0) \in W_0$ then

$$\lim_{l \to \infty} y_l(k) = \pm s_{i_0}(k) \quad \forall k = 1, 2, ...,$$

where $y_l(k) = \mathbf{w}(l)^T\mathbf{x}(k)$ and $i_0 = i(\mathbf{w}(0))$.

(c) The rate of convergence in (b) is of order $p - 1$.

4. Examples and remarks for practical implementation

Below we consider some examples, for concrete values of p.

1) $p = 4$. Then

$$\varphi_4(\mathbf{w}) = \mathbf{cum}_4(\mathbf{w}^T\mathbf{x}) = E\{(\mathbf{w}^T\mathbf{x})^4\} - 3(E\{(\mathbf{w}^T\mathbf{x})^2\})^2$$

and

$$\varphi_4'(\mathbf{w}) = 4E\{(\mathbf{w}^T\mathbf{x})^3\mathbf{x}\} - 12E\{(\mathbf{w}^T\mathbf{x})^2\}E\{(\mathbf{x}\mathbf{x}^T)\}\mathbf{w}.$$

We note that if the standard prewhitening is made (i.e. $E\{\mathbf{x}\mathbf{x}^T\} = \mathbf{I}_n$, $\mathbf{A}$ is orthogonal), the algorithm (12) recovers the fixed-point algorithm of Hyvärinen and Oja, i.e.

$$\mathbf{w}(l+1) = \frac{E\{(\mathbf{w}(l)^T\mathbf{x})^3\mathbf{x}\} - 3\mathbf{w}(l)}{\|E\{(\mathbf{w}(l)^T\mathbf{x})^3\mathbf{x}\} - 3\mathbf{w}(l)\|}$$

2) $p = 6$. Then

$$\begin{aligned}\varphi_6(\mathbf{w}) &= E\{(\mathbf{w}^T\mathbf{x})^6\} - 15E\{(\mathbf{w}^T\mathbf{x})^2\}E\{(\mathbf{w}^T\mathbf{x})^4\}\\ &\quad -10(E\{(\mathbf{w}^T\mathbf{x})^3\})^2 + 30(E\{(\mathbf{w}^T\mathbf{x})^2\})^3,\end{aligned}$$

$$\begin{aligned}\varphi_6'(\mathbf{w}) &= 6E\{(\mathbf{w}^T\mathbf{x})^5\mathbf{x}\} - 30E\{\mathbf{x}\mathbf{x}^T\}\mathbf{w}E\{(\mathbf{w}^T\mathbf{x})^4\}\\ &\quad -60E\{(\mathbf{w}^T\mathbf{x})^2\}E\{(\mathbf{w}^T\mathbf{x})^3\mathbf{x}\}\\ &\quad -60E\{(\mathbf{w}^T\mathbf{x})^3\}E\{(\mathbf{w}^T\mathbf{x})^2\mathbf{x}\}\\ &\quad +180(E\{(\mathbf{w}^T\mathbf{x})^2\})^2E\{\mathbf{x}\mathbf{x}^T\}\mathbf{w}.\end{aligned}$$

If, in addition, $E\{\mathbf{x}\mathbf{x}^T\} = \mathbf{I}$, then

$$\begin{aligned}\varphi_6'(\mathbf{w}) &= 6E\{(\mathbf{w}^T\mathbf{x})^5\mathbf{x}\} - 30\mathbf{w}E\{(\mathbf{w}^T\mathbf{x})^4\}\\ &\quad -60E\{(\mathbf{w}^T\mathbf{x})^3\mathbf{x}\}\\ &\quad -60E\{(\mathbf{w}^T\mathbf{x})^3\}E\{(\mathbf{w}^T\mathbf{x})^2\mathbf{x}\}\\ &\quad +180\mathbf{w}.\end{aligned}$$

We have implemented the algorithm (12) in the case of six order cumulants, using a deflation procedure. We generated 50 sparse source signals and run the algorithm after random mixture. The number of iterations for extracting all signals one by one using six order cumulants was 313, and the number of iterations using fourth order cumulants was 339. But the computational time is bigger in the case of six order cumulants. This can be explained by the computational time needed for calculation of the six order cumulants, since they have more complex structure.

We should mention also that the successful extraction of components depends of their statistical independence. In case when they are not so independent, the extraction is problematic and depends on the initial condition.

5. Combining second and fourth order statistics

In this section we consider an unified model of source signals and additive noise, which is white of order 2 and 4. We assume that all source signals are uncorrelated of order 2 and 4, as some of the source signals (we don't know which) are white of order 4 but colored of order 2 (for instance colored Gaussian signals) and the rest are white of order 2 and colored of order 4. We introduce a new sufficient condition for separation (see condition **DCF**(P) below) stating that the sources have different autocorrelation functions or different cumulant functions of fourth order over a given set P of time delays. This condition can be considered as a generalization of those ones described in [5] and [17] (for second order statistics) and used in [4].

The second and fourth order statistics is used in [11] and [14] by a joint diagonalization procedure of covariance and cumulant matrices, combining SOBI and JADE algorithms described in [2] and [3] respectively. Our idea below is to use cumulant matrices involving time delays.

Define a covariance matrix of the sensor (resp. source) signals by

$$\mathbf{R_x}(p) = E\{\mathbf{x}\mathbf{x}_p^T\}, \text{ (resp. } \mathbf{R_s}(p) = E\{\mathbf{s}\mathbf{s}_p^T\}), \tag{13}$$

where E is the mathematical expectation, $\mathbf{x}_p = \mathbf{x}(k-p), \mathbf{x} = \mathbf{x}(k), \mathbf{s}_p = \mathbf{s}(k-p), \mathbf{s} = \mathbf{s}(k)$, and the symmetric matrix $\tilde{\mathbf{R}}_\mathbf{x}(p)$ by

$$\tilde{\mathbf{R}}_\mathbf{x}(p) = \frac{1}{2}\left(\mathbf{R_x}(p) + \mathbf{R_x}(p)^T\right).$$

Define a fourth order cumulant matrix $\mathbf{C}^{2,2}_{\mathbf{x},\mathbf{x}_p}$ of the sensor signals by

$$\mathbf{C}^{2,2}_{\mathbf{x},\mathbf{x}_p} = E\{\mathbf{x}\mathbf{x}^T(\mathbf{x}_p^T\mathbf{x}_p)\} - E\{\mathbf{x}\mathbf{x}^T\}\mathrm{tr}E\{\mathbf{x}_p\mathbf{x}_p^T\} - 2E\{\mathbf{x}\mathbf{x}_p^T\}E\{\mathbf{x}_p\mathbf{x}^T\},$$

and the symmetric matrix $\tilde{\mathbf{C}}^{2,2}_{\mathbf{x},\mathbf{x}_p}$ by

$$\tilde{\mathbf{C}}^{2,2}_{\mathbf{x},\mathbf{x}_p} = \frac{1}{2}\left(\mathbf{C}^{2,2}_{\mathbf{x},\mathbf{x}_p} + (\mathbf{C}^{2,2}_{\mathbf{x},\mathbf{x}_p})^T\right).$$

Note that $\mathbf{C}^{2,2}_{\mathbf{x},\mathbf{x}_p}$ (resp. $\mathbf{R_x}(p)$) is symmetric, if $\mathbf{C}^{2,2}_{\mathbf{s},\mathbf{s}_p}$ (resp. $\mathbf{R_s}(p)$) is a diagonal matrix, but in order to avoid the effect of computational errors (which could destroy the symmetricity), we use $\tilde{\mathbf{C}}^{2,2}_{\mathbf{x},\mathbf{x}_p}$ (resp. $\tilde{\mathbf{R}}_\mathbf{x}(p)$).

It is easy to see that the (i,j)-th element of $\mathbf{C}^{2,2}_{\mathbf{x},\mathbf{x}_p}$ is

$$C^{2,2}_{\mathbf{x},\mathbf{x}_p}(i,j) = \sum_{l=1}^{n} \mathbf{cum}\{x_i(k), x_j(k), x_l(k-p), x_l(k-p)\}$$

(see [3] for more general cumulant matrices).

Similarly we define analogous matrices $\mathbf{C}^{2,2}_{\mathbf{s},\mathbf{s}_p}$ and $\tilde{\mathbf{C}}^{2,2}_{\mathbf{s},\mathbf{s}_p}$ for the source signals $\mathbf{s}(k)$. Recall that a signal s is white of order 2 (resp. white of order 4) if

$$E\{s(k)s(k-p)\} = 0, \quad \forall p \geq 1$$

(resp. $\mathbf{cum}\{s(k-p_1), s(k-p_2), s(k-p_3), s(k-p_4)\} = 0$

for every $p_i \geq 1, i = 1, ..., 4$) (see [15]).

In a linear data model (1), if the noise $\mathbf{n}$ is white of order 4, and the mixing matrix $\mathbf{A}$ is orthogonal, then the time-delayed cumulant matrices of the observation vector $\mathbf{x}(k)$ for any $p \neq 0$ satisfy

$$\tilde{\mathbf{C}}^{2,2}_{\mathbf{x},\mathbf{x}_p} = \mathbf{A}\tilde{\mathbf{C}}^{2,2}_{\mathbf{s},\mathbf{s}_p}\mathbf{A}^T. \quad (14)$$

If $\mathbf{n}$ is white of order 2, then

$$\tilde{\mathbf{R}}_{\mathbf{x}}(p) = \mathbf{A}\tilde{\mathbf{R}}_{\mathbf{s}}(p)\mathbf{A}^T. \quad (15)$$

The representations (14) and (15) gives the idea how we can achieve separation by combination of fourth and second order statistics: we jointly diagonalize matrices of type $\tilde{\mathbf{C}}^{2,2}_{\mathbf{x},\mathbf{x}_i}$ for $i = 1, ..., q$ and matrices of type $\tilde{\mathbf{R}}_{\mathbf{x}}(j)$ for $j = 1, ..., p$. This is our algorithm called SFOBI_pq, which we use in the sequel.

In order this procedure to be successful, the following condition, called **DCF**(P) (different cumulant functions) must be satisfied:

$$\begin{aligned}
&\forall i, j \neq i \, \exists l_{i,j} \in \{1, ..., L\} : \\
&\text{either } E\{s_i(t)s_i(t-p_{l_{i,j}})\} \neq E\{s_j(t)s_j(t-p_{l_{i,j}})\} \\
&\text{or } \mathbf{cum}_{s_i}(p_{l_{i,j}}) \neq \mathbf{cum}_{s_j}(p_{l_{i,j}}),
\end{aligned}$$

i.e. the sources have different autocorrelation or cumulant functions of fourth order on a given set $P = \{p_1, ..., p_L\}$ of time delays. For a detailed description of this condition and its necessity, see [10].

6. Experiment with EEG data for detection of eye movements (Experimental comparison of methods employing second order statistics, fourth order statistics and their combination)

Our experimental setup consisted of a NeuroScan electroencephalographic (EEG) system (Neurosoft Inc.) equipped with a 64-channel QuickCap using Ag-AgCl-type electrodes. The electrical currents from the scalp were amplified by two SynAmps amplifiers and digitally recorded using SCAN acquision software at a sampling rate of 1000 Hz. The biosignals were bandpass-filtered and electrical noise was removed outside the range 0.05-200 Hz. Electrode impedance was lower than 5 KOhm for all channels except for channel C4. The subjects were seated in a dark shielded room and the images were projected by two Marquee-type CRT projectors on the left side of a screen located 2.8m frontally.

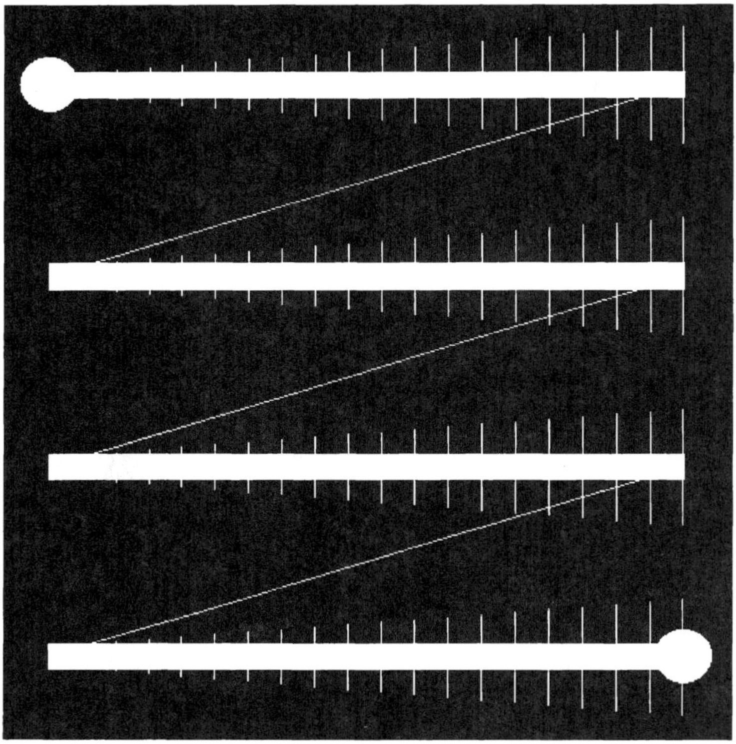

Figure 3.1. Eye movement pattern projected on a screen in front of the subjects.

At the beginning of each eye movement trial, a pattern image appeared on the screen, consisting of 4 white horizontal thick lines on a black background (see Fig. 3.1). The end of each horizontal thick line was connected to the beginning of the next horizontal line by a thin diagonal line. After the pattern appeared on the screen, the subject had to perform the following task: fixate a small circular gaze point on the left side of the first line for 0.5 seconds, then move the gaze smoothly along each horizontal line and cross diagonally to the next one. At the end of the fourth horizontal eye movement, fixate the gaze point again for 0.5 seconds on the small circle on the right side of the line, then press a button to indicate that the task is finished. Pressing the button caused the pattern to disappear and after a 7-second break a short warning sound called for attention. Additional 3 seconds were given for concentration before the next trial pattern appeared and the procedure was repeated. After artifact rejection and segmentation procedures (8192ms epochs), 19 channels corresponding to the international 10/20 system were extracted from the original data set for better visualization. The segmented eye movement data was processed using independent component analysis (ICA). We employed three methods in order to separate the ocular electrical potentials from the evoked cortical activity: second order method (algorithm EVD2), fourth order method (algorithm JADETD) and a combined 2nd and 4th order statistics method (algorithm SFOBI_pq) described in Section 5. These algorithms are implemented within the framework of the ICALAB for Signal Processing software package [7].

7. Results

Our experiments showed that eye movement information could be extracted from the electroencephalogram without using dedicated horizontal and vertical oculographic electrodes.

Single-trial eye-movement EEG data (Fig. 3.2) was processed using three component extraction algorithms. We compared the performance of the combined 2nd and 4th order SFOBI_pq algorithm with the 2nd-order EVD2 and the 4th order JADETD algorithms by setting two goals: 1) extraction of eye movement data in single-trial EEG recordings, and 2) separation of the horizontal and vertical components within the 4-line eye movements.

The extracted components showed that the 4th-order JADETD algorithm (Fig. 3.3c) is not able to extract any of the relatively slow eye movement components, while the combined 2nd- and 4th-order SFOBI_pq (Fig. 3.3a), as well as the 2nd-order EVD2 (Fig. 3.3b) both completed the task of extraction. In SFOBI_pq, the number of time-delayed covariance matrices p had to be chosen

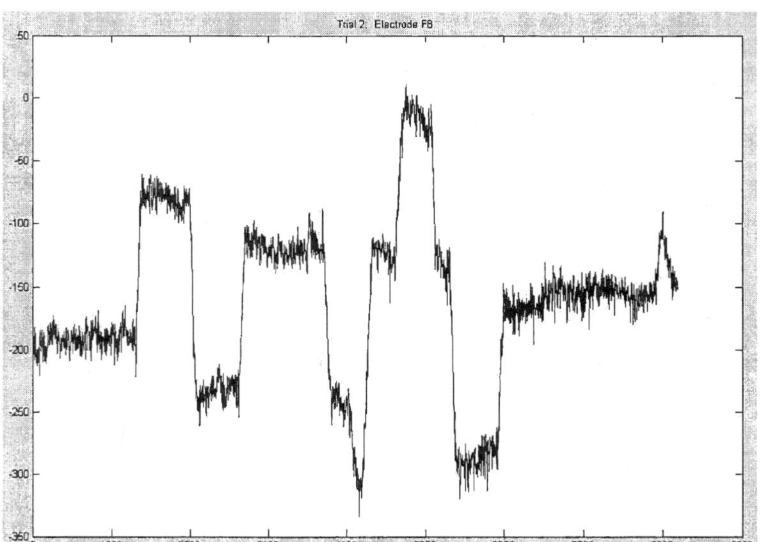

Figure 3.2. Single-trial eye-movement EEG data recorded from electrode F8 which is located near the right eye. This trial exhibits two deviations from the intended movement-rest routine - between the 3rd and 4th line scans and at the end of the trial during the resting period.

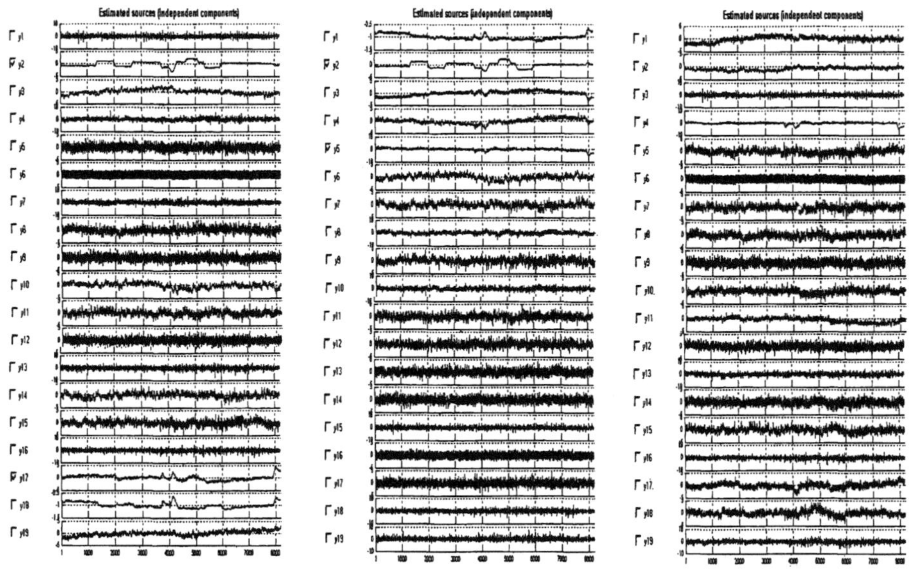

Figure 3.3. Independent components showing the performance of the three compared extraction algorithms: (a) combined 2nd and 4th order SFOBI_pq (p=10, q=100), (b) 2nd-order EVD2, (c) 4th order JADETD.

to be greater than 1 and smaller than the number of time-delayed cumulant matrices q. Below we examined the latter two algorithms in order to compare their performance in the second goal - separating the horizontal and vertical eye movement components. We selected two components generated by each of the SFOBI_pq and EVD2 packages using the following criteria: 1) the first component must have increased signal-to-noise ratio during horizontal eye movements, while the second component must complement the first one for vertical and diagonal movements, 2) if there are two similar components available, then the following selection criteria were applied by projecting components back into signal space (signal reconstruction): 2A) if several similar components result also in similar signal changes after reconstruction then the component with the maximum impact (projected peak amplitude) was chosen, 2B) adding or removing a second component should not change or distort significantly the reconstructed signal from the first one.

Fig. 3.4 shows reconstructed EEG signals after selecting and deflating both components for each SFOBI_pq and EVD2. We found that after reconstructing the horizontal eye movement component the signal in right lateral channel F8 (according to the International 10/20 EEG system) was strongest, while reconstructing the vertical/diagonal component influenced most strongly right anterior channel FP2. Although both algorithms matched the combined eye

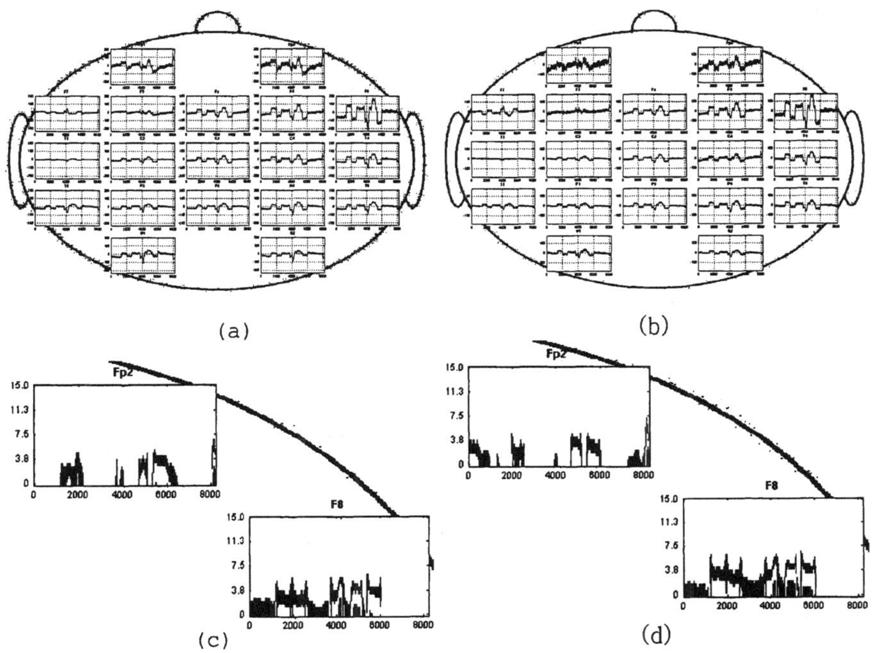

Figure 3.4. Reconstruction of eye movement signals from 2 components which represent the electrical activity originating from the combined horizontal and vertical motion of the eyes. (a) signal reconstruction for all channels using combined 2nd and 4th order SFOBI_pq (p=10, q=100), (b) signal reconstruction using 2nd-order EVD2. Time-frequency analysis spectrograms in (c) and (d) show the frequency domain changes in time for channels F8 and FP2 which demonstrated strongest sensitivity for horizontal and diagonal eye movements correspondingly. EVD2 exhibited spurious vertical eye movement activity in channel FP2 at the beginning and near the end of the trial.

movement, EVD2 exhibited spurious vertical eye movement activity in channel FP2 at the beginning and near the end of the trial.

Fig. 3.5 and Fig. 3.6 illustrate the separate reconstruction of the horizontal and vertical eye movement components for SFOBI_pq and EVD2. SFOBI_pq successfully removed a short vertical saccade at the end of the trial from the reconstructed horizontal movement data, while EVD2 failed (Fig.5). Furthermore, in the vertical component projection (Fig. 3.6), SFOBI_pq showed an advantage in extracting all 3 weak vertical eye movement potentials, while EVD2 entirely failed to detect the 3rd diagonal movement and exhibited false positives in the frequency domain at the beginning and before the end of the trial.

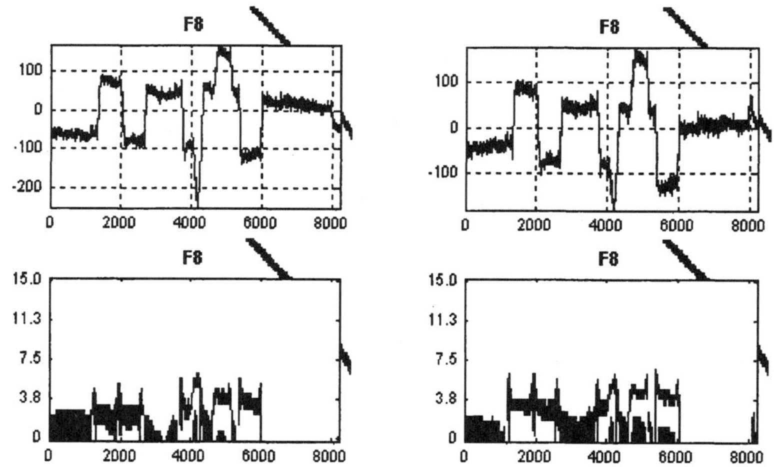

(a) (b)

Figure 3.5. Reconstruction of left-to-right horizontal eye movement signal in right lateral electrode F8 from a single component. (a) reconstruction using combined 2nd and 4th order SFOBI_pq (p=10, q=100), (b) reconstruction using 2nd-order EVD2. Although for horizontal eye movements the performance of both algorithms was relatively similar, EVD2 incorporated also the sharp vertical movement at the end of the trial and failed to separate it from the 'horizontal' component.

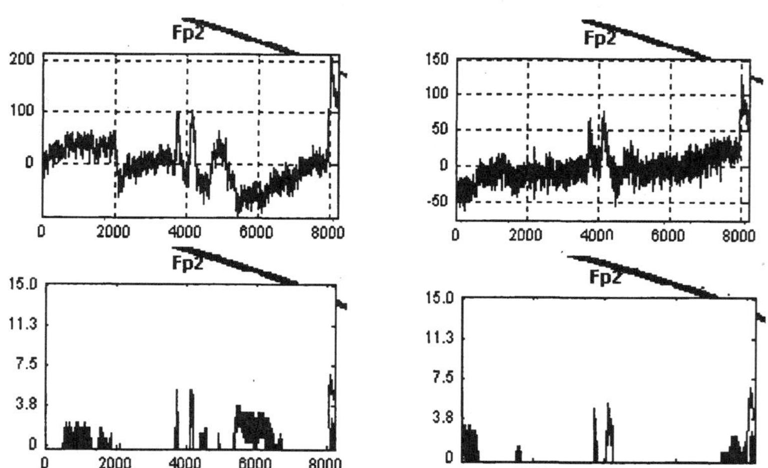

(a) (b)

Figure 3.6. Reconstruction of diagonal eye movement signal in right anterior electrode FP2 from a single component. (a) reconstruction using combined 2nd and 4th order SFOBI_pq (p=10, q=100), (b) reconstruction using 2nd-order EVD2. SFOBI_pq showed an advantage in extracting all 3 weak vertical eye movement potentials, while EVD2 failed to detect the 3rd diagonal movement and exhibited false positives in the frequency domain at the beginning and just before the end of the trial.

8. Conclusion

We presented a mathematical justification of maximization of cumulants method for ICA and give generalizations of: 1) the fixed point algorithm of Hyvärinen and Oja for high order cumulants, and 2) cumulants method, combining second and fourth order statistics by joint approximate diagonalization of covariant and cumuant matrices depending on time delays. In our eye movement EEG experiments, results indicated that the proposed combined 2nd- and 4th-order algorithm exhibited a better performance than the 2nd-order and the 4th-order algorithms in extracting and separating properly the strong horizontal movement signals from the weak diagonal movement potentials.

References

[1] V.M. Alekseev, V.M. Tihomirov, and Fomin S.V. *Optimal Control.* Nauka, 1979. (in Russian).

[2] A. Belouchrani, K.A. Meraim, J.-P. Cardoso and E. Moulines A blind source separation technique using second order statistics. *IEEE Transaction on Signal Processing*, 45(2):434–444, February 1997.

[3] J.-F. Cardoso. High-order contrasts for independent component analysis. *Neural Computation*, 11(1):157–192, January 1999.

[4] C. Chang, Z. Ding, S. F. Yau, and F. H. Y. Chan. A matrix-pencil approach to blind separation of colored nonstationary signals. *IEEE Transactions on Signal Processing*, 48(3):900–907, March 2000.

[5] C. Chang, S. F. Yau, P. Kwok, F. H. Y. Chan, and F. K. Lam. Uncorrelated and component analysis for blind source separation. *Circuits Systems and Signal Processing*, 18(3):225–239, 1999.

[6] A. Cichocki and S. Amari. *Adaptive Blind Signal and Image Processing.* John Wiley, Chichester, 2002.

[7] A. Cichocki, S. Amari, C. Siwek, et al. ICALAB toolboxes. `http : //www.bsp.brain.riken.go.jp/ICALAB/`.

[8] N. Delfosse and P. Loubaton. Adaptive blind separation of independent sources: a deflation approach. *Signal Processing*, 45:59–83, 1995.

[9] P. G. Georgiev and A. Cichocki. On some new ideas and algorithms for independent component analysis. In *Proceedings Workshop on Nonlinear Analysis and Convex Analysis 2001*, pages 19–30, Kyoto, Japan, 2002. Research Institute for Mathematical Sciences.

[10] P. G. Georgiev and A. Cichocki. Robust Blind Source Separation utilizing second and fourth order statistics. In *Proceedings Intern. Conf. Artificial Neural Networks–ICANN, Madrid, Spain, 2002. Lect. Notes Comp. Sci. 2415*, pages 1162–1167.

[11] I. Gorodnitsky and A. Belouchrani. Joint cumulant and correlation based signal separation with application of eeg data analysis. In *Proc. 3-rd Int. Conf. on Independent Component Analysis and Signal Separation, San Diego, California*, pages 475–480, 2001.

[12] A. Hyvärinen, J. Karhunen, and E. Oja. *Independent Component Analysis*. John Wiley & Sons, 2001.

[13] A. Hyvärinen and E. Oja. A fast fixed-point algorithm for independent component analysis. *Neural Computation*, 9:1483–1492, 1997.

[14] M. Klajman and A. G. Constantinides. A weighted mixed statistics algorithm for blind source separation. In *Proc. EUSIPCO'02*, Toulousse, France, 2002.

[15] C. L. Nikias and A. Petropulu. *Higher-order spectra analysis. A nonlinear signal processing framework*. Signal Processing. Prentice Hall, 1993.

[16] A. N. Shiryaev. *Probability*. Graduate Texts in Mathematics. Springer, 1996.

[17] L. Tong, R. Liu, V. C. Soon, and Y. F. Huang. Indeterminacy and identifiability of blind identification. *IEEE Transactions on Circuits and Systems*, 38(5):499–509, May 1991.

Chapter 4

ON A NEW QUANTIZATION IN COMPLEX SYSTEMS

Victor Korotkikh
Faculty of Informatics and Communication
Central Queensland University
Mackay, Queensland 4740, Australia
v.korotkich@cqu.edu.au

Galina Korotkikh
Faculty of Informatics and Communication
Central Queensland University
Mackay, Queensland 4740, Australia
g.korotkich@cqu.edu.au

Abstract Significant progress has been made recently in epileptic seizure prediction. Further advances may be confronted by the absence of quantitative measures of structures in complex systems and challenged because the laws of the brain probably still far away from our knowledge. In this context the paper suggests to seek the answers within an irreducible description of complex systems. A special type of processes, the formations of integer relations, is proposed to describe and quantize complex systems in an irreducible way. As an approximation this results in a new complexity measure expressed as a trace of variance-covariance matrix. A connection between the measure and measures being developed for structures in complex systems is presented.

Keywords: brain disorders, complex systems, measures of complexity, formation processes of integer relations, traces of variance-covariance matrix.

1. Introduction

Methods of complex systems open new ways in dealing with brain disorders [1]. In particular, recently significant progress has been made in epileptic seizure prediction by using short-term maximum Lyapunov exponents [2, 3]. At the same time further advances in this direction may be confronted by the absence of quantitative measures of structures in dynamics of complex systems [4]. Moreover, the situation is challenged because the laws of the brain may be still far away from our knowledge and experience.

In this context the paper suggests to seek the answers within an irreducible description of complex systems. A special type of processes, i.e., the formations of integer relations [5], is used to describe and quantize complex systems. A key feature of these processes is that they are irreducible. This property takes place because the formations of integer relations are completely controlled by arithmetic and have the integers as the ultimate building blocks.

The formation processes are unusual in the sense that they have not been observed in physical experiments. Information about a formation process of integer relations can be obtained from a system of linear equations describing a nonlocal correlation between sequences [5, 6]. These known results are briefly given in sections 2 and 3 to introduce the formation processes.

A geometrical interpretation of the integer relations formations is presented in section 4. The interpretation even allows to imagine the integer relations as some sort of particles, called integer particles [7].

In section 5 it is shown how to characterize and quantify structures of a complex system in terms of the formations of integer relations. As an approximation this results in a new complexity measure expressed in terms of a trace of variance-covariance matrix. The interval of the trace is used to measure the complexity of the structures and by this way the complexity of the states. Computational experiments suggest a possible quantization of the interval into separate regions with similar structures.

A connection between the measure and measures being developed for structures in complex systems [8–10] is presented. This may suggest a new perspective to describe and quantize complex systems in terms of the formations of integer relations.

Main results of the paper are summarized in conclusions.

2. A Nonlocal Correlation between Sequences

In this section a notion of nonlocal correlation between two different sequences is presented [6]. The correlation produces a hierarchical structure consisting of parts of the sequences. It is possible to express the nonlocal correlation as a system of linear equations in integers. The system of equations gives information about a special type of processes, i.e., the formations of integer relations [5].

Let I be an integer alphabet and

$$I_n = \{s = s_1 ... s_n,\ s_i \in I,\ i = 1, ..., n\}$$

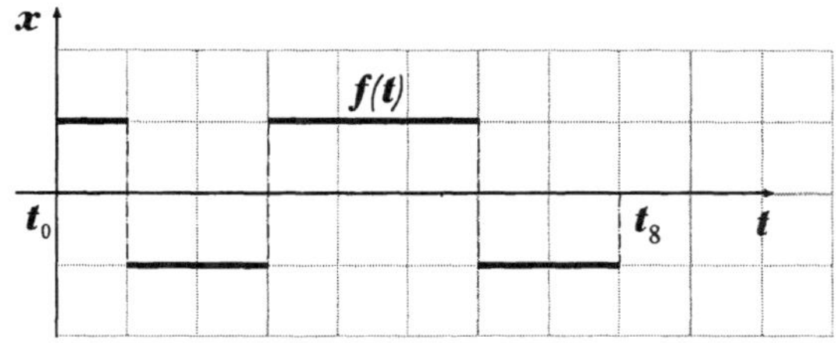

Figure 4.1. Graph of a function $f = \rho_{m\varepsilon\delta}(s),\ s = +1 - 1 - 1 + 1 + 1 + 1 - 1 - 1$.

be the set of all sequences of length $n \geq 2$ with symbols in I. Let $\delta > 0$ and $\varepsilon > 0$ be respective spacings of a space-time lattice (δ, ε) in $1 + 1$ dimensions.

Let $W_{\varepsilon\delta}([t_m, t_{m+n}])$ be a class of piecewise constant functions such that a function f of the class is constant on $(t_{i-1}, t_i]$, $i = m+1, ..., m+n$ and equals

$$f(t_m) = s_1\delta,\ f(t) = s_i\delta,\ t \in (t_{i-1}, t_i],$$

$$i = m+1, ..., m+n,\ t_i = i\varepsilon,\ i = m, ..., m+n,$$

where m is an integer and $s_i,\ i = 1, ..., n$ are real numbers. Sequence $s = s_1...s_n$ is called a code of the function f, denoted $s = c(f)$. Figure 4.1 shows a function f such that

$$f \in W_{11}([t_0, t_8]),\quad c(f) = +1 - 1 - 1 + 1 + 1 + 1 - 1 - 1.$$

Let $\rho_{m\varepsilon\delta} : s \to f$ be a mapping that associates a sequence $s \in I_n$ with a function $f \in W_{\varepsilon\delta}[t_m, t_{m+n}]$, denoted $f = \rho_{m\varepsilon\delta}(s)$, such that $c(f) = s$ and whose kth integral satisfies $f^{[k]}(t_m) = 0,\ k = 1, 2, \dots$.

We characterize a sequence $s = s_1...s_n \in I_n$ by successive integrals

$$s \Longrightarrow f^{[1]}(t), f^{[2]}(t), ..., f^{[k]}(t), ...$$

of a function $f = \rho_{m\varepsilon\delta}(s)$. Figures 4.2 and 4.3 illustrate the characterization of a sequence

$$s = +1 - 1 - 1 + 1 + 1 + 1 - 1 - 1$$

by showing $f^{[1]}(t), f^{[2]}(t)$, where $n = 8, m = 0, \varepsilon = 1, \delta = 1$.

We consider a nonlocal correlation between two different sequences

$$s = s_1...s_n \in I_n,\quad s' = s'_1...s'_n \in I_n$$

that results in the following correlation between $C(s, s') \geq 1$ integrals

$$f^{[k]}(t_{m+n}) = g^{[k]}(t_{m+n}),\quad k = 1, ..., C(s, s') \tag{1}$$

of functions $f = \rho_{m\varepsilon\delta}(s), g = \rho_{m\varepsilon\delta}(s')$. The correlation between the sequences s, s' does not extend to the $(C(s, s') + 1)$th integrals of the functions

$$f^{[C(s,s')+1]}(t_{m+n}) \neq g^{[C(s,s')+1]}(t_{m+n}). \tag{2}$$

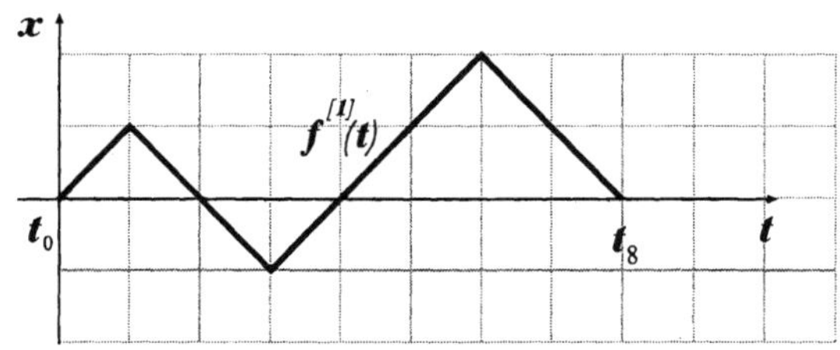

Figure 4.2. Graph of the first integral $f^{[1]}$ of function f in Figure 4.1.

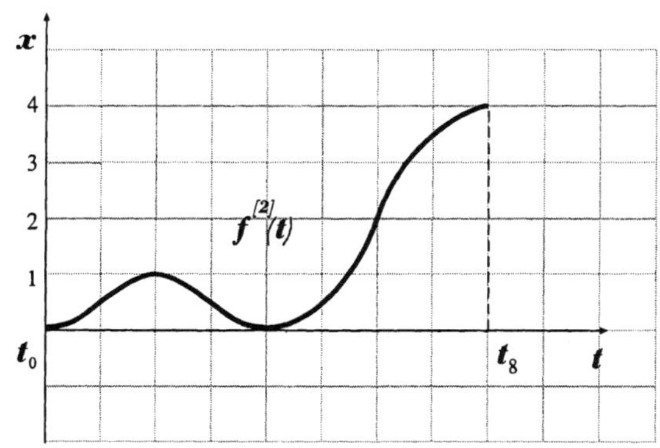

Figure 4.3. Graph of the second integral $f^{[2]}$ of function f in Figure 4.1.

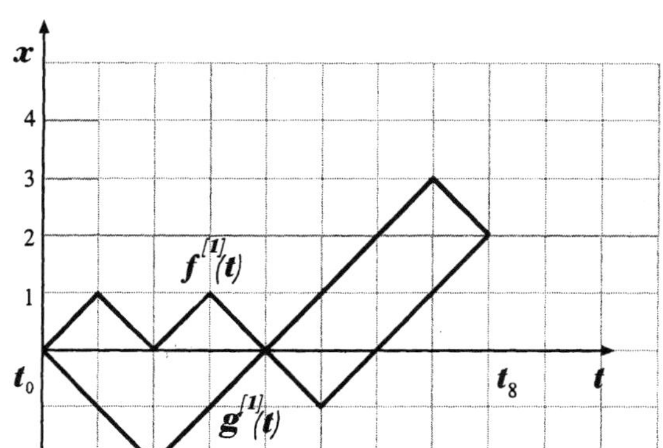

Figure 4.4. First integrals $f^{[1]}$ and $g^{[1]}$ show the correlation. Starting at t_0 the integrals move differently in the interval but come together at t_8. As the move from one step to another is restricted, the integrals are correlated to meet at the right end.

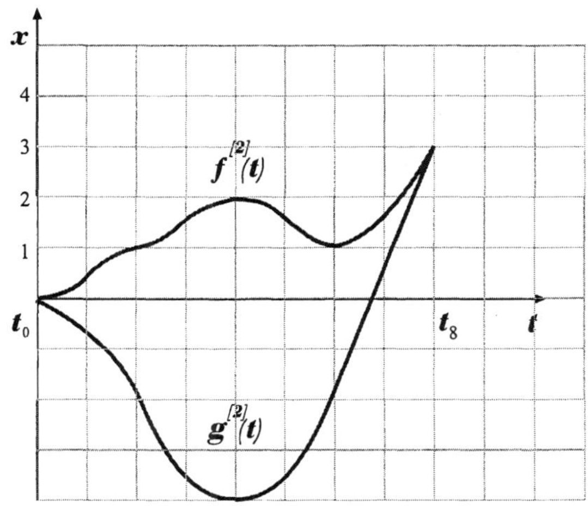

Figure 4.5. Second integrals $f^{[2]}$ and $g^{[2]}$ show the correlation. Starting at t_0 the integrals move differently in the interval but the correlation makes them to come together at t_8.

Figures 4.4 and 4.5 present the situation for sequences

$$s = +1-1+1-1-1+1+1+1, \; s' = -1-1+1+1+1+1+1-1.$$

The correlation between the sequences s, s' results in the correlation between their first and second integrals

$$f^{[1]}(t_8) = g^{[1]}(t_8), \quad f^{[2]}(t_8) = g^{[2]}(t_8)$$

of functions

$$f = \rho_{011}(s), \quad g = \rho_{011}(s').$$

An integer code series [11] expresses the nonlocal correlation in terms of a system of linear equations in integers and makes it possible to find formation processes of integer relations [5].

3. The Formations of Integer Relations: Processes with Integer Particles

The integer code series expresses an integral of a piecewise constant function

$$f \in W_{\varepsilon\delta}([t_m, t_{m+n}])$$

in terms of the code $c(f)$, powers of integers and combinatorial coefficients [11].

Integer Code Series (V. Korotkikh, 1988). *Let* $f \in W_{\varepsilon\delta}([t_m, t_{m+n}])$ *be a piecewise constant function such that* $c(f) = s_1 ... s_n$. *Then the kth integral*

$f^{[k]}$, $k \geq 1$ *of the function* f *at a point* $t_{m+l}, l = 1, ..., n$ *can be given by*

$$f^{[k]}(t_{m+l}) = \sum_{i=0}^{k-1} \alpha_{kmi}((m+l)^i s_1 + ... + (m+1)^i s_l)\varepsilon^k \delta$$

$$+ \sum_{i=1}^{k} \beta_{kli} f^{[i]}(t_m)\varepsilon^{k-i}, \tag{3}$$

where $\alpha_{kmi}, \beta_{kli}$, $i = 1, ..., k$ *are combinatorial coefficients.*

The integer code series (3) for sequences

$$s = s_1...s_n \in I_n, \quad s' = s'_1...s'_n \in I_n,$$

such that $C(s, s') \geq 1$ gives

$$f^{[k]}(t_{m+n}) = \sum_{i=0}^{k-1} \alpha_{kmi}((m+n)^i s_1 + (m+n-1)^i s_2 + ... + (m+1)^i s_n)\varepsilon^k \delta,$$

$$g^{[k]}(t_{m+n}) = \sum_{i=0}^{k-1} \alpha_{kmi}((m+n)^i s'_1 + (m+n-1)^i s'_2 + ... + (m+1)^i s'_n)\varepsilon^k \delta,$$

where $f = \rho_{m\varepsilon\delta}(s)$, $g = \rho_{m\varepsilon\delta}(s')$ and $k = 1, ..., C(s, s')$.

It is proved in [5] by using (3) that if for sequences

$$s = s_1...s_n \in I_n, \quad s' = s'_1...s'_n \in I_n$$

we have $C(s, s') \geq 1$, then $C(s, s') \leq n$ and condition (1) reduces to a system of $C(s, s')$ equations

$$(m+n)^0(s_1 - s'_1) + ... + (m+1)^0(s_n - s'_n) = 0$$

$$. \quad . \quad . \quad . \quad . \quad .$$

$$(m+n)^{C(s,s')-1}(s_1 - s'_1) + ... + (m+1)^{C(s,s')-1}(s_n - s'_n) = 0 \tag{4}$$

while condition (2) results in an inequality

$$(m+n)^{C(s,s')}(s_1 - s'_1) + ... + (m+1)^{C(s,s')}(s_n - s'_n) \neq 0. \tag{5}$$

Notice, that if $C(s, s') = n$ then system (4) appears with the matrix

$$\begin{pmatrix} (m+n)^0 & (m+n-1)^0 & ... & (m+1)^0 \\ (m+n)^1 & (m+n-1)^1 & ... & (m+1)^1 \\ . & ... & & . \\ (m+n)^{n-1} & (m+n-1)^{n-1} & ... & (m+1)^{n-1} \end{pmatrix}$$

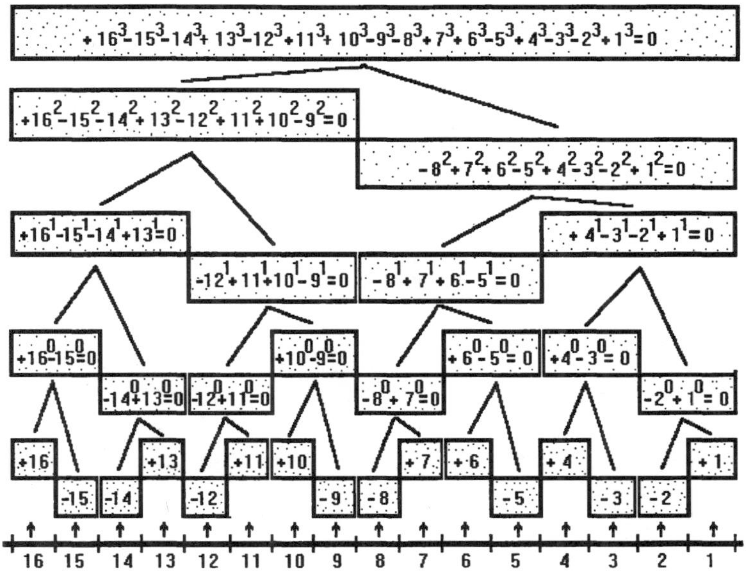

Figure 4.6. Represented in this form system (6) appears as the formation of integer relations with integers 16, ..., 1 as ultimate building blocks. In the formation all integer relations have the same organizing principle.

whose determinant is Vandermonde one.

For example, for the Prouhet-Thue-Morse sequences (starting with +1 and −1) we have $C(s, s') = 4$ and (4) becomes a system of integer relations ($n = 16, m = 0$ and factor 2 is ignored for clarity)

$$+16^0 - 15^0 - 14^0 + 13^0 - 12^0 + 11^0 + 10^0 - 9^0 - 8^0 + 7^0 + 6^0 - 5^0 + 4^0 - 3^0 - 2^0 + 1^0 = 0$$

$$+16^1 - 15^1 - 14^1 + 13^1 - 12^1 + 11^1 + 10^1 - 9^1 - 8^1 + 7^1 + 6^1 - 5^1 + 4^1 - 3^1 - 2^1 + 1^1 = 0$$

$$+16^2 - 15^2 - 14^2 + 13^2 - 12^2 + 11^2 + 10^2 - 9^2 - 8^2 + 7^2 + 6^2 - 5^2 + 4^2 - 3^2 - 2^2 + 1^2 = 0$$

$$+16^3 - 15^3 - 14^3 + 13^3 - 12^3 + 11^3 + 10^3 - 9^3 - 8^3 + 7^3 + 6^3 - 5^3 + 4^3 - 3^3 - 2^3 + 1^3 = 0 \quad (6)$$

whereas (5) does not follow the character in (6)

$$+16^4 - 15^4 - 14^4 + 13^4 - 12^4 + 11^4 + 10^4 - 9^4 - 8^4 + 7^4 + 6^4 - 5^4 + 4^4 - 3^4 - 2^4 + 1^4 \neq 0. \quad (7)$$

The analysis of system (4) can identify important features [5]:

- integers can be seen as ultimate building blocks of integer relations of the system (see Figure 4.7 as illustration),
- there may be more integer relations than the system can show,
- there are relationships between the integer relations, which can be described in terms of an organizing principle.

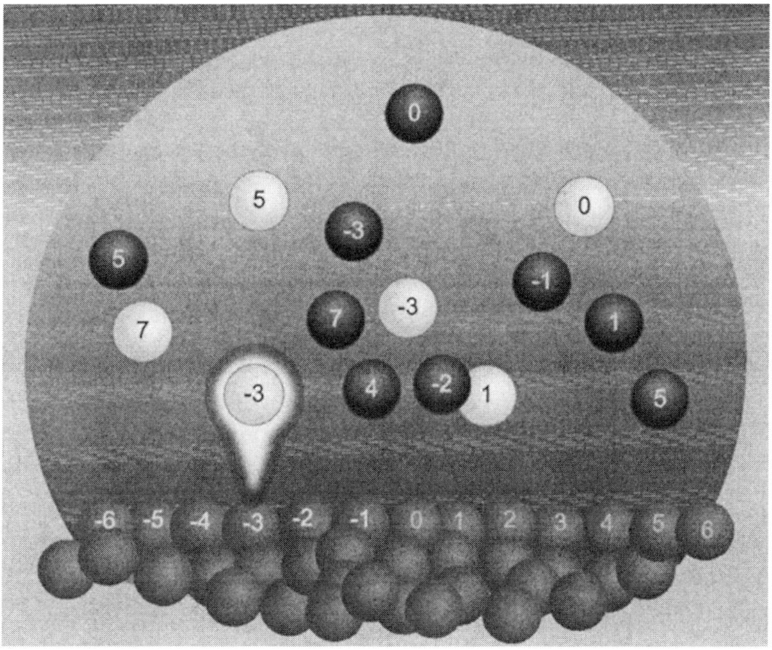

Figure 4.7. In the figure integers seem like particles that can combine and form composite particles. There can be positive integer particles (shown in black) and negative integer particles (shown in white). For example, we can see a positive integer particle $+(-3)$ and a negative integer particle $-(-3)$.

These features suggest to talk about some type of process connected with the system (4) and even imagine the integer relations as some sort of particles, called integer particles [7]. The system of linear equation (4) does not allow to see the process as in physical experiments, but makes it possible to observe the process implicitly by giving information about its objects and their relationships.

In particular, system (4) can be associated with a hierarchical set

$$WR(s, s', n, m, I_n)$$

of $C(s, s') \geq 1$ levels whose elements of level $k = 1, ..., C(s, s')$ are integer relations of the form

$$A_1 d_1^{k-1} + ... + A_l d_l^{k-1} = 0,$$

where $A_i, d_i,\ i = 1, ..., l$ are integers $d_i > d_{i+1},\ i = 1, ..., l-1$ and k is the power of $d_i,\ i = 1, ..., l$ [5]. It is also interpreted that elements of level $k = 2, ..., C(s, s')$ of the set are formed from elements of level $(k-1)$ by the organizing principle. The formation organizing principle can be described as follows. If $r \geq 1$ integer relations

$$A_{i1} d_{i1}^{k-1} + ... + A_{il(i)} d_{il(i)}^{k-1} = 0 \tag{8}$$

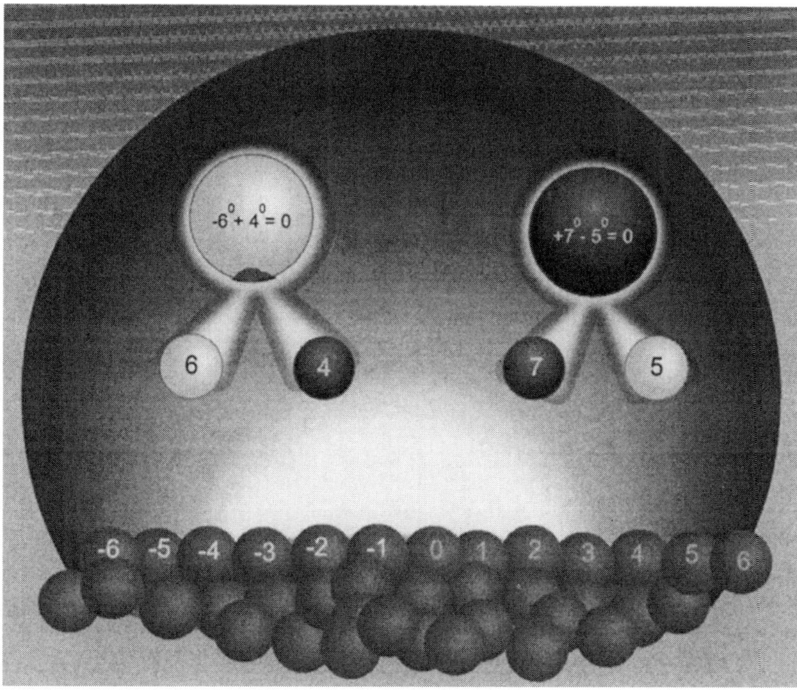

Figure 4.8. The figure illustrates the formation of integer relations from integers, which both are viewed as integer particles. Negative integer particle 6 and positive integer particle 4 under the "interaction" produce $-(6)^0+4^0$, which is a composite integer particle made of them because $-(6)^0+4^0=0$.

of level $k = 1, ..., C(s, s') - 1$, with relation i, $i = 1, ..., r$ containing $l(i)$ terms, satisfy

$$\sum_{i=1}^{r} A_{i1}d_{i1}^{k} + ... + A_{il(i)}d_{il(i)}^{k} = 0, \tag{9}$$

then integer relation (9) is formed from integer relations (8). Figure 4.8 illustrates the formations of integer relations of the first level from integers.

In Figure 4.6 showing (6) as the formation, we can see that integers are generated on the zero level and then form integer relations of the first level. These integer relations in turn form integer relations of the second level. The formation continues to the fourth level, where the integer relation, because of (7), can not alone form an element of the next level.

Thus, the nonlocal correlation (1) between sequences

$$s = s_1...s_n \in I_n, \quad s' = s_1'...s_n' \in I_n$$

can be described in terms of nonlocal correlations between integers (4) and is realized as the formation of integer relations, also denoted by $WR(s, s', n, m, I_n)$.

A formation process $WR(s, s', n, m, I_n)$ is a complexity type process. The level to which the formation process progresses is $C(s, s')$ and is called a structural complexity of sequence s with respect to sequence s' [5].

The formation processes of integer relations are irreducible because they are completely controlled by arithmetic and have the integers as the ultimate building blocks [5].

The nonlocal correlation considered in terms of the integer code series gives us a characterization of sequences. The characterization recognizes in the sequence a hierarchical structure consisting of its parts and shows how it is formed. In this structure even the most distant parts may have the relationship. The structure is sensitive to the order of the components $s_i, i = 1, \ldots, n$ in the sequence $s = s_1 \ldots s_n$ and even a minor variation in the order may lead to its significant changes.

This property is quite different from how entropy characterizes complex systems. In particular, the entropy of a system described by a discrete probability distribution function as a sequence $s = s_1 ... s_n$ is given by

$$H = -\sum_{i=1}^{n} \frac{s_i}{\bar{\mathrm{s}}} log_2 \frac{s_i}{\bar{\mathrm{s}}}, \tag{10}$$

where $\bar{\mathrm{s}} = \sum_{i=1}^{n} s_i,\ s_i \geq 0, i = 1, ..., n$ and the system is quantified according to the value of (10).

A permutation of the components $s_i, i = 1, ..., n$ of the sequence s does not change the entropy. Thus, entropy cannot sense the structures which the nonlocal correlation deals with. Entropy may stand in some sense for the average of the structures of the sequence's permutations.

Many quantitative measures of complex systems are based on entropy. For example Kolmogorov-Sinai entropy of a complex system could find an expression as

$$H_{KS} = \sum_{\lambda_i > 0} \lambda_i,$$

where λ_i is a positive Lyapunov exponent. The comment about entropy may explain why the measures of complex systems are insensitive to structures arising in dynamics of complex systems [4].

4. Geometrization of the Integer Relations Formations

The concepts of integer and integer relations are an integral part of our mental equipment. However, they are too abstract to give us insight what we think the formation is all about. It is difficult to imagine how integer relations can be really formed from other integer relations, because there are no ways to visualize how this may actually happen.

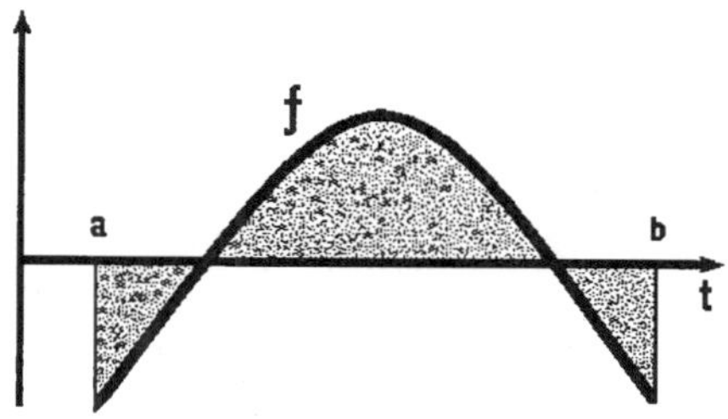

Figure 4.9. The pattern (shaded) of a function f is a two-dimensional geometrical object and can be visualized.

But is it possible to see how the integer relations look like and how they form into each other? To answer the question a new understanding of the integer relations as geometrical objects that can form into each other is developed by using a notion of integer pattern [5]. The notion is defined by using a notion of function pattern (see Figure 4.9). Integer patterns are function patterns with special properties [5].

It is proved by using (3) that there exists an isomorphism $\psi_{\varepsilon\delta}$ between elements of a set $WR(s, s', n, m, I_n)$, i.e., integer relations, and elements of a set $WP_{\varepsilon\delta}(s, s', n, m, I_n)$, i.e., integer patterns of integrals $f^{[k]}$, $k = 1, ..., C(s, s')$, where

$$f = \rho_{m\varepsilon\delta}(s - s') \in W_{\varepsilon\delta}([t_m, t_{m+n}]),$$

and between their formations as well [5]. The isomorphism $\psi_{\varepsilon\delta}$ gives a geometrical meaning to the integer relations and their formations in terms of integer patterns and their formations. In particular, the integer relations and their formation processes can be quantitatively described by corresponding integer patterns. The organizing principle appears as function integration.

The formation of integer relations and the corresponding formation of integer patterns are shown together in Figure 4.10. We can see an integer relation by a corresponding integer pattern and measure by its area. The formation of integer patterns in Figure 4.10 reveals self-similarity, local and nonlocal symmetries.

5. Quantization of States of a Complex System as the Formations of Integer Relations

Currently, there is no systematic way to characterize and quantify the states of a complex system. In the absence mainly two approaches are used: two-point correlation functions and dynamical invariants such as Lyapunov exponents and fractal dimensions [4]. Both approaches are limited in applications [4, 12].

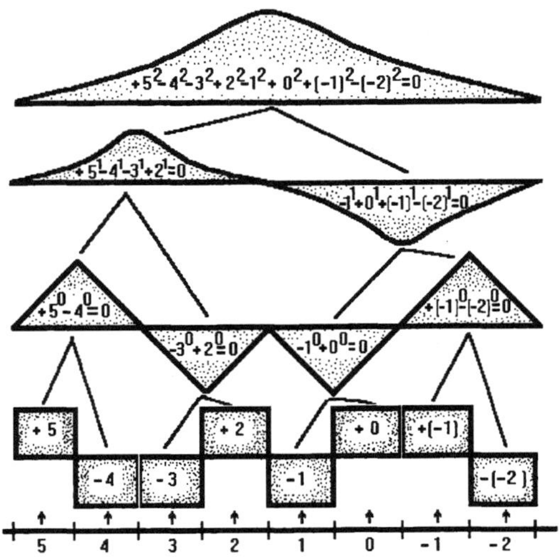

Figure 4.10. The formation of integer relations and the corresponding formation of integer patterns (sketched) are pictured together to show their unified character. The figure allows us to see how the integer relations form into each other. An integer relation can be measured by the area of a corresponding integer pattern and the nonlocal correlation by the areas of all integer patterns.

A quantization characterizes a state of a complex system by an element of a structure, whose elements are distinguishable and to some extend comparable. A quantitative description of an element of the structure may be used to measure a corresponding state. The more the quantization becomes adequate with reality, the more we question why the structure is so special.

The *structure* incorporating the formations of integer relations into one whole has not only rich properties for a quantization of complex systems but also, if consistent with experimental facts, will not allow a deeper explanatory base. In this quantization a state of a complex system can be described by a formation process of integer relations and measured by corresponding integer patterns. One state may be compared with another state by using corresponding formation processes.

The comparison is based on the order of the formations of integer relations and may be interpreted in terms of complexity. A formation process produces integer relations of one level from integer relations of the previous one. Therefore, integer relations of level $k \geq 1$ may be seen more complex than integer relations of level $k - 1$, because they are made of integer relations of the lower level. Consequently, one state of a complex system described by integer relations of level k and higher may be seen more complex than another state described by integer relations of levels lower than level k.

Let information about a complex system be given by $N \geq 2$ time series, for example by a number of EEG recordings, each describing a basic part of the system. Then a state of the system is specified by a $N \times n$ matrix

$$S = \{s_{ij}\}_{i=1,\ldots,N;\ j=1,\ldots,n},$$

where a sequence $\bar{s}_i = s_{i1} \ldots s_{in} \in I_n,\ i = 1, \ldots, N$ is a time series of part i observed at instants $t_j = j\varepsilon,\ j = 1, ..., n$. Let $\mathbf{S}(I_n, N)$ be the set of all states of the complex system.

The states of the complex system can be characterized and quantified by the formations of integer relations. In particular, a state $S \in \mathbf{S}(I_n, N)$ can be described by a set of formation processes

$$WR(S, n, m) = \bigcup_{i=1}^{N} \bigcup_{j=1}^{N} WR(\bar{s}_i, \bar{s}_j, n, m, I_n),$$

specified by a $N \times N$ matrix

$$C(S) = \{C(\bar{s}_i, \bar{s}_j)\}_{i,j=1,\ldots,N},$$

where $C(\bar{s}_i, \bar{s}_j)$ is the structural complexity of sequence $\bar{s}_i, i = 1, \ldots, N$ with respect to sequence $\bar{s}_j, j = 1, \ldots, N$.

Let us explain the description of a state $S \in \mathbf{S}(I_n, N)$ of the complex system. A formation process $WR(\bar{s}_i, \bar{s}_j, n, m, I_n)$ describes the part of the system composed of basic parts i and j and their relationship. Integer relations produced by the formations $WR(S, n, m)$ may be involved in formation processes resulting in integer relations of higher levels.

The following rule applies and sets a correspondence between integer relations and parts of the complex system. Parts can form a composed part of the system if their corresponding integer relations can form integer relations of higher levels corresponding to the composed part of the system. As a result the correlation length, denoted ξ_C, in the complex system increases, because the parts become correlated in the composite part. The composite part in its turn may combine with other parts of the complex system. In particular, integer relations produced by one formation and integer relations produced by another formation of the same level may form integer relations of a higher level. These integer relations together with integer relations produced on this level by a different formation may form integer relations of a more higher level.

Therefore, starting with integer relations produced by $WR(S, n, m)$ the formation processes may compose integer relations of higher levels. This corresponds to the construction of a structure of the state S of the complex system. In the structure each part of the complex system is presented as a result of the formation from smaller parts and possibly as a block in the formations of larger parts of the system. The geometrical interpretation of the integer relations can be used to measure the relationships between the parts of the complex system.

For example, consider a state $S \in \mathbf{S}(B_4, 3)$ of a complex system consisting of three basic parts $\bar{s}, \bar{s}', \bar{s}''$

$$S = \begin{pmatrix} +1 & -1 & -1 & +1 \\ -1 & +1 & +1 & -1 \\ +1 & -1 & +1 & -1 \end{pmatrix} \tag{11}$$

where

$$\bar{s} = +1 - 1 - 1 + 1, \quad \bar{s}' = -1 + 1 + 1 - 1, \quad \bar{s}'' = +1 - 1 + 1 - 1$$

and B_4 is the set of binary sequences of length 4. For simplicity the parts and the sequences are denoted in the same way.

A formation process $WR(s, s', 4, 0, B_4)$ results in an integer relation

$$\boxed{+4^1 - 3^1 - 2^1 + 1^1 = 0} \tag{12}$$

of the second level and formation processes

$$WR(s', s'', 4, 0, B_4), \quad WR(s'', s, 4, 0, B_4)$$

give integer relations

$$\boxed{-4^0 + 3^0 = 0}$$

$$\boxed{+2^0 - 1^0 = 0} \tag{13}$$

of the first level accordingly. Integer relations (13) form an integer relation

$$\boxed{-4^1 + 3^1 + 2^1 - 1^1 = 0} \tag{14}$$

of the second level. In a proper representation integer relations (12) and (14) become

$$\boxed{+8^1 - 7^1 - 6^1 + 5^1 = 0}$$

$$\boxed{-4^1 + 3^1 + 2^1 - 1^1 = 0}$$

and form an integer relation

$$\boxed{+8^2 - 7^2 - 6^2 + 5^2 - 4^2 + 3^2 + 2^2 - 1^2 = 0}$$

of the third level. These formation processes of integer relations correspond to the formation of a structure of the complex system (see Figure 4.11).

In particular, it can be seen that the composite part at the right on the second level is made of two parts, which are also composite. Each of these two parts is made from basic parts and both contain the basic part $\bar{s}''$, in one case with the relationship directed to $\bar{s}''$ and from $\bar{s}''$ in the other. Thus, it may be said

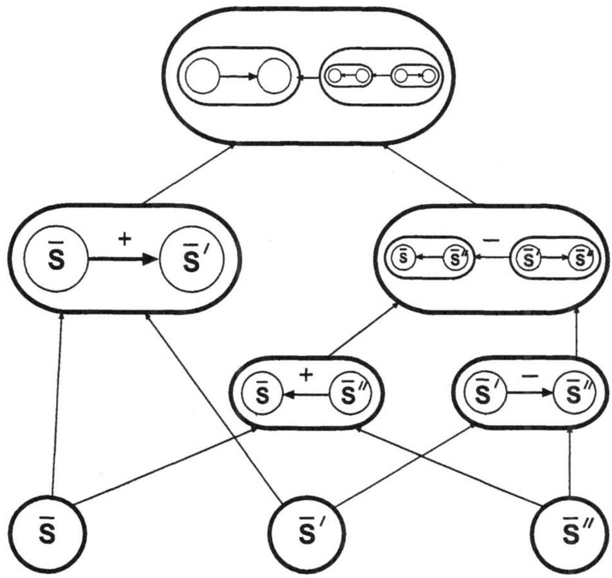

Figure 4.11. The figure shows the formation of a structure of a state (11). The process is hierarchical with the resulting structure at the top. A part at level k, $1 \leq k \leq 3$ can be positive or negative and is composed of parts from level $k-1$.

that in the composite part the basic part $\bar{s}''$ is represented in a superposition. When observations of the composite part could be imagined to find part $\bar{s}''$ in one way or the other, it is tempting even to interpret this situation in quantum mechanical terms.

There is a wide range of possible structures of the states. They correspond to the structures constructed by the formations of integer relations. From one side there can be no integer relations produced by $WR(S, n, m)$ for a state $S \in \mathbf{S}(I_n, N)$. In this case the basic parts do not form any composite part and we have a structure of minimum complexity. From another side there can be integer relations produced by $WR(S, n, m)$ for a state $S \in \mathbf{S}(I_n, N)$ in an optimal way so that resulting integer relations of the highest possible levels can be formed. In this case we have a structure of maximum complexity.

Importantly, the *structure* of the structures of the states is well enough defined. It is known that an element of this *super-structure*, i.e., the structure of a state, corresponds to a formation process of integer relations and that the elements are organized in it by the formations of integer relations. This information may be helpful in understanding phase transitions of a complex system. In a phase transition the structure of a complex system changes from one to another.

A question arises: how to characterize and quantify these structures of a complex system in an efficient way. An approach is proposed to use a set of variance-covariance matrices derived from state $S \in \mathbf{S}(I_n, N)$ and employ traces of these matrices [6].

In this section we consider the first approximation of the approach, which uses the variance-covariance matrix

$$V(S) = \{V(\bar{s}_i, \bar{s}_j)\}_{i,j=1,\ldots,N},$$

of the state $S \in \mathbf{S}(I_n, N)$, where

$$V(\bar{s}_i, \bar{s}_j) = \frac{E(\bar{s}_i \bar{s}_j) - E(\bar{s}_i)E(\bar{s}_j)}{\sqrt{\sigma^2(\bar{s}_i)\sigma^2(\bar{s}_j)}},$$

is the linear correlation coefficient between $\bar{s}_i$ and $\bar{s}_j$, $\sigma^2(\bar{s}_i), \sigma^2(\bar{s}_j)$ are the variances of $\bar{s}_i$ and $\bar{s}_j$, and $E(\,)$ denotes a time average over a period considered.

Let

$$Spec(V(S)) = (\lambda_1, \ldots, \lambda_N)$$

be the eigenvalue spectrum of the variance-covariance matrix $V(S)$ of a state $S \in \mathbf{S}(I_n, N)$ and

$$tr(V^2(S)) = \lambda_1^2 + \cdots + \lambda_N^2$$

be the quadratic trace of the variance-covariance matrix. Let $\mathbf{V}(I_n, N)$ be the set of all variance-covariance matrices $V(S)$ of the states $S \in \mathbf{S}(I_n, N)$.

As the first approximation it is suggested to characterize and quantify the structures of a complex system described in terms of the formations of integer relations by using the quadratic trace $tr(V^2(S))$ of the variance-covariance matrix $V(S)$, $S \in \mathbf{S}(I_n, N)$ [6]. For this purpose it is useful to consider two extreme cases, i.e., one of minimum complexity structures and the other of maximum complexity structures.

Minimum complexity structure corresponds to a state $S \in \mathbf{S}(I_n, N)$ when all sequences of the state are the same, i.e., all basic parts $i = 1, \ldots, N$ behave in the same manner. In this case the complex system can be described by one basic part, because it behaves as the complex system itself. There are no integer relations produced by $WR(S, n, m)$ and basic parts do not form a composite part. In this case any combination of parts behaves as the parts themselves.

The variance-covariance matrix $V(S)$ of such a state $S \in \mathbf{S}(I_n, N)$ is

$$V(S) = V_{min} = \begin{pmatrix} 1 & 1 & \ldots & 1 & 1 \\ 1 & 1 & \ldots & 1 & 1 \\ . & . & \ldots & . & . \\ 1 & 1 & \ldots & 1 & 1 \\ 1 & 1 & \ldots & 1 & 1 \end{pmatrix}, \quad Spec(V_{min}) = (N, 0, \ldots, 0).$$

Maximum complexity structure is considered with the help of a variance-covariance matrix V_{max} whose coefficients, except the diagonal ones, are zero

$$V_{max} = \begin{pmatrix} 1 & 0 & \dots & 0 & 0 \\ 0 & 1 & \dots & 0 & 0 \\ . & . & \dots & . & . \\ 0 & 0 & \dots & 1 & 0 \\ 0 & 0 & \dots & 0 & 1 \end{pmatrix}, \quad Spec(V_{max}) = (1, ..., 1)$$

and is realized for a state $S \in \mathbf{S}(I_n, N)$ such that $V(S) = V_{max}$.

From these two situations we have from one side

$$tr(V_{min}^2) = \max_{V \in \mathbf{V}(I_n,N)} tr(V^2) = \max_{V \in \mathbf{V}(I_n,N)} \sum_{i=1}^{N} \lambda_i^2 =$$

$$N^2 + \underbrace{0^2 + \cdots + 0^2}_{N-1} = N^2 \tag{15}$$

subject to

$$tr(V) = \sum_{i=1}^{N} \lambda_i = N, \quad \lambda_i \geq 0, \ i = 1, \dots, N$$

and another side

$$tr(V_{max}^2) = \min_{V \in \mathbf{V}(I_n,N)} tr(V^2) = \min_{V \in \mathbf{V}(I_n,N)} \sum_{i=1}^{N} \lambda_i^2 = N \tag{16}$$

subject to

$$tr(V) = \sum_{i=1}^{N} \lambda_i = N, \quad \lambda_i \geq 0, \ i = 1, \dots, N.$$

Combining (15) and (16) for the variance-covariance matrix $V(S) \in \mathbf{V}(I_n, N)$ of a state $S \in \mathbf{S}(I_n, N)$ we obtain

$$N \leq tr(V^2(S)) \leq N^2. \tag{17}$$

By using (17) we get an interval for the quadratic trace $tr(V^2(S))$ of a state $S \in \mathbf{S}(I_n, N)$

$$1 \leq \frac{N^2}{tr(V^2(S))} \leq N, \tag{18}$$

which is considered to characterize and quantify the structures of the states. In particular, for a state $S \in \mathbf{S}(I_n, N)$ let

$$K_{FIR}(S) = \frac{N^2}{tr(V^2(S))},$$

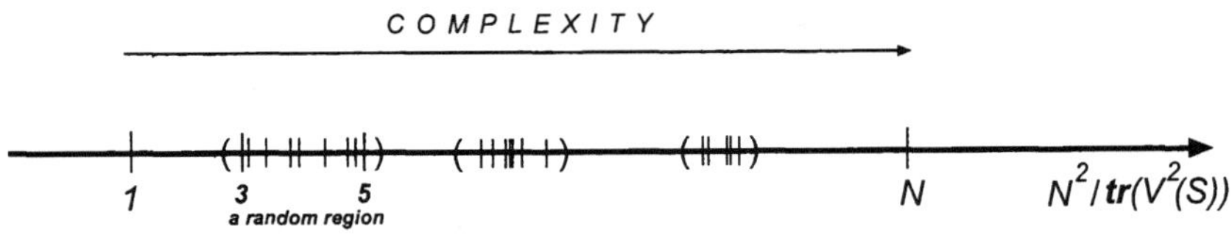

Figure 4.12. The interval is used to measure the complexity of the integer relations structures and by this way the complexity of the states. Computational experiments suggest a possible quantization of the interval into separate regions with similar structures. A region representing random states is specifically shown.

where FIR stands to show the connection with the formations of integer relations. We have for K_{FIR} from (18)

$$1 \leq K_{FIR}(S) \leq N. \tag{19}$$

The quantity $K_{FIRS}(S)$, called a structural complexity of the state, is suggested to measure the complexity of the structure of the state S and by this way the complexity of the state S itself. The structural complexity $K_{FIR}(S)$ increases from the minimum at the left end of the interval to the maximum bounded by the right end of the interval. Maximum structural complexity of a complex system, according to (19), could increase as the number of its basic parts N becomes bigger.

The integer relations structures have a discrete nature, i.e., the integer relations are organized into hierarchical levels. This suggests a possible quantization of the quadratic trace interval (18) into separate regions with similar structures. This quantization of the interval would be useful in a quantization of the structures and thus in identifying qualitative changes or phase transitions in complex systems.

In this context computational experiments have been made. They give some evidence to suggest that such a quantization of the interval (18) may be in place. In particular, random states have been studied by generating random binary sequences to construct from them random walk functions as the rows of the matrix S. The computations show that the structural complexity $K_{FIR}(S)$ of such a random state S independent of N belongs to a region

$$3 \preceq K_{FIR}(S) \preceq 5$$

as long as N and n are bigger enough. Figure 4.12 illustrates some results of the computational experiments.

The structural complexity K_{FIR} is a new way to characterize complex systems. It is important to relate K_{FIR} with other similar means for dealing with complex systems. A connection between the structural complexity K_{FIR} and the number of KLD (Karhunen-Loeve decomposition) modes D_{KLD} can be identified. It is suggested in many disciplines to use D_{KLD} to measure the complexity of spatiotemporal data in large, high-dimensional, nontransient, driven-dissipative systems [8–10].

We briefly present D_{KLD} by using [4] to show the connection. The Karhunen-Loeve decomposition is a statistical method for compressing spatiotemporal data by finding the largest linear subspace that contains substantial variations of the data.

Let $u(t, x)$ be a one-dimensional zero-mean field on a spatial interval whose values are measured on a finite space-time lattice of n uniformly sampled time points $t_i = i\varepsilon,\ i = 1, \ldots, n$ and of N uniformly sampled spacial points $x_j = j\delta,\ j = 1, \ldots, N$. The measurement is analogous to EEG recordings when spacial points are interpreted as electrode sites. A $n \times N$ rectangular data matrix

$$A = \{A_{ij} = u(t_i, x_j)\}_{i=1,\ldots,n;\ j=1,\ldots,N},$$

can be defined from which a $N \times N$ symmetric positive semidefinite scatter matrix $M = A^T A$ can be calculated, where A^T denotes the matrix transpose of A. The scatter matrix can be diagonalized to obtain its nonnegative eigenvalues $\lambda_i^2,\ i = 1, \ldots, N$, which can be further decreasingly ordered

$$\lambda_1^2 \geq \lambda_2^2 \geq \ldots \geq \lambda_N^2 \geq 0.$$

A positive integer

$$D_{KLD} = \max\{p : \frac{\sum_{i=1}^{p} \lambda_i^2}{\sum_{i=1}^{N} \lambda_i^2} \leq r\} \tag{20}$$

is introduced to characterize the eigenvalues. It represents the largest number of KLD modes p needed to capture some specified fraction $r \leq 1$ of the total variance $\sum_{i=1}^{N} \lambda_i^2$ of the data.

We can find from (20) that D_{KLD} similar to K_{FIR} in (19) belongs to the interval

$$1 \leq D_{KLD} \leq N, \tag{21}$$

which D_{KLD} quantizes uniformly.

It can be seen that if $K_{FIR} = 1$ then $D_{KLD} = 1$ for $r \leq 1$ and if $K_{FIR} = N$ then $D_{KLD} = N$ when r belongs to some interval.

The connection allows to apply the structural complexity K_{FIR} for complex systems being studied by D_{KLD}. More importantly, it reveals D_{KLD} in the context of the K_{FIR} conceptual framework, i.e., the formations of integer relations. In particular, this shows where complexity interpretation of D_{KLD} may come from and that a different quantization of the interval (21), not the uniform one of D_{KLD}, could be more adequate for complex systems.

6. Conclusions

Significant progress has been made recently in epileptic seizure prediction [2, 3]. Further advances in dealing with brain disorders may be confronted by the absence of quantitative measures of structures in complex systems [4]. Moreover, the situation is challenged because the laws of the brain may be still far away from our knowledge and experience.

In this context the paper suggests to seek the answers within an irreducible description of complex systems. In particular, a special type of processes, i.e., the formations of integer relations [5], is proposed to describe and quantize complex systems in an irreducible way. As an approximation this results in a new complexity measure expressed in terms of a trace of variance-covariance matrix.

A connection between the measure and measures being developed for structures in complex systems is presented. This may suggest a new perspective to describe and quantize complex systems in terms of the formations of integer relations.

References

[1] J. Milton and P. Jung (Eds.), *Epilepsy as a Dynamic Disease*, Springer-Verlag, New York, 2002.

[2] L. Iasemidis, J. Sackellares, P. Pardalos, D. Shiau, W. Chaovalitwongse and P. Carney, *Epileptic Seizure Prediction*, in Proceedings of Quantitative Neurosciences, University of Florida, February 5-7, 2003, p. 19.

[3] W. Chaovalitwongse, P. Pardalos, J. Sackellares, L. Iasemidis, D. Shiau, W. Chaovalitwongse and P. Carney, *Nonlinear Dynamics and Global Optimization in EEG with Applications for Prediction of Eliptic Seizures*, in Proceedings of Quantitative Neurosciences, University of Florida, February 5-7, 2003, p. 23.

[4] S. Zolde and H. Greenside, *Karhunen-Loeve Decomposition of Extensive Chaos*, Physical Review Letters, **78**, 9, 1997, pp. 1687-1690.

[5] V. Korotkich, *A Mathematical Structure for Emergent Computation*, Kluwer Academic Publishers, Dordrecht/Boston/London, 1999.

[6] G. Korotkikh and V. Korotkikh, *On the Role of Nonlocal Correlations in Optimization*, in Optimization and Industry: New Frontiers, P. Pardalos and V. Korotkikh (Eds.), Kluwer Academic Publishers, Dordrecht/Boston/London, 2003, pp. 181-220.

[7] V. Korotkikh, *An Approach to the Mathematical Theory of Perception-Based Information*, in Fuzzy Partial Differential Equations and Relational Equations: Reservoir Characterization and Modelling, M. Nikravesh, L.

Zadeh and V. Korotkikh (Eds.), Physica-Verlag, Heidelberg/New York, 2003, pp. 80-115 (in press).

[8] L. Sirovich and A. Deane, *J. Fluid Mech.* **222**, 251, (1991).

[9] S. Ciliberto and B. Nicolaenko, *Europhys. Lett.* **14**, 303, (1991).

[10] R. Vautard and M. Ghil, *Physica*(Amsterdam) **35D**, 395, (1989).

[11] V. Korotkich, *Integer Code Series: Applications in Dynamic Systems and Complexity*, The Computing Center of Russian Academy of Sciences, Moscow, 1993.

[12] H. Abarbanel, R. Brown and L. Tsimring, *Rev. Mod. Phys.* **65**, 1331, (1993).

Chapter 5

THE SEIZURE PREDICTION CHARACTERISTIC

A new terminology and assessment criterion for epileptic seizure prediction methods

T. Maiwald

FDM, Center for Data Analysis and Modeling, University of Freiburg, Eckerstr. 1, 79104 Freiburg, Germany

maiwald@fdm.uni-freiburg.de

M. Winterhalder

FDM, Center for Data Analysis and Modeling, University of Freiburg, Eckerstr. 1, 79104 Freiburg, Germany

winterhm@fdm.uni-freiburg.de

H.U. Voss

FDM, Center for Data Analysis and Modeling, University of Freiburg, Eckerstr. 1, 79104 Freiburg, Germany

hv@fdm.uni-freiburg.de

R. Aschenbrenner-Scheibe

Epilepsy Center, University Hospital of Freiburg, Breisacher Str. 64, 79106 Freiburg, Germany

aschenbs@nz.ukl.uni-freiburg.de

A. Schulze-Bonhage

Epilepsy Center, University Hospital of Freiburg, Breisacher Str. 64, 79106 Freiburg, Germany

schulzeb@nz.ukl.uni-freiburg.de

J. Timmer
FDM, Center for Data Analysis and Modeling, University of Freiburg, Eckerstr. 1, 79104 Freiburg, Germany
jeti@fdm.uni-freiburg.de

Abstract Epilepsy is characterized by the sudden occurrence of seizures disturbing the perception or behavior of epileptic patients. Several prediction methods have claimed to be able to predict seizures based on EEG-recordings minutes in advance, which opens up new approaches to treat the disease. However, the term seizure prediction is not unequivocally defined and different assessment criteria for prediction methods exist which impedes the comparison between methods. Moreover, only little attention is paid to the dependency between sensitivity and false prediction rate. We address these shortcomings and introduce a terminology and assessment criterion for seizure prediction methods based on statistical and clinical considerations: the seizure prediction characteristic.

Keywords: Epilepsy, seizure prediction, seizure prediction characteristic, sensitivity, false prediction rate

1. Introduction

In his work "On the Sacred Disease" Hippocrates (400 B.C.) refutes a curse as origin of epileptic seizures but suggests a brain disorder. Today we know that during seizures abnormal electrical discharges in the brain impairs its functioning. Astonishingly, five in hundred persons suffer from one epileptic seizure during life and one in hundred persons have recurrent seizures. Anticonvulsive drugs or surgery therapies cannot help 12% of the patients. These patients have to cope with the incessant uncertainty of arising seizures. A method to predict seizures could improve therapeutic strategies dramatically. So far, several seizure prediction methods based on the analysis of intracranial and scalp EEG have been suggested with promising results which are summarized in table 5.1. For a review see [Litt and Lehnertz, 2002] and [Litt and Echauz, 2002].

A seizure prediction method could be applied clinically by triggering an intervention system which is able to control an arising seizure, for example via the administration of strong anticonvulsive drugs into the epileptic focus or electrical stimulation of the vagus nerve. A simple intervention would be the warning of the patient. He could avoid dangerous situations like a busy street or a swimming pool.

Figure 5.1 provides an example of how a seizure prediction method works. A mathematical algorithm extracts a "feature" from the EEG recording. Once this feature crosses a specific threshold level, an alarm is triggered. A comparison of interictal periods far away from any seizure and pre-ictal periods resulting in seizure onset leads to the choice of a suitable threshold value. In this case,

Author		*Seizure prediction method*			
# Patients	*# Seizures*	*Interictal*	*MPT*	*Sensitivity*	*FPR*
[Martinerie et al., 1998]		correlation density			
11	19	0 h	2.64 min	89%	/
[Osorio et al., 1998]		based on frequency analysis			
13	125	34 h	0.25 min	92%	0 FP/h
[Lehnertz and Elger, 1998]		effective correlation dimension			
16	16	5.2-34.7 h	11.5 min	94%	0 FP/h
[Le Van Quyen et al., 1999]		dynamical similarity index			
13	23	0 h	5.75 min	83%	/
[Le Van Quyen et al., 2000]		dynamical similarity index			
9	17	0 h	4.45 min	94%	/
[Iasemidis et al., 2001]		Lyapunov exponent			
5	58	0 h	49.1 min	91%	/
[Le Van Quyen et al., 2001]		dynamical similarity index (surface EEG)			
23	26	0 h	7 min	96%	/
[Lehnertz et al., 2001]		effective correlation dimension			
59	95	> 115 h	19 min	47%	0 FP/h
[Jerger et al., 2001]		seven different prediction methods			
4	12	0 h	1 - 3 min	/	/
[De Clercq et al., 2002]		similarity index, correlation dimension (surface EEG)			
12	12	0 h	/	0%	/
[Schindler et al., 2002]		leaky integrator and fire unit (surface EEG)			
7	15	144 h	4-330 min	100%	> 0.014 FP/h
[Navarro et al., 2002]		dynamical similarity index			
11	41	12-60 h	7.54 min	83%	0.3 FP/h
[Mormann et al., 2003]		phase coherence, linear cross correlation			
10	14	15 h	86/ 102 min	86%	0 FP/h

Table 5.1. Achievements of seizure prediction methods developed to date. Listed are the number of patients and seizures investigated, the total duration of interictal EEG data for calculation of the false prediction rate, the mean prediction time (MPT), sensitivity, and the rate of false predictions per hour (FPR). False prediction rates of 0 FP/h mean that no false prediction occurred for the investigated EEG data. All but three studies were done with intracranial EEG data.

lower threshold values correspond to a higher sensitivity since more seizures can be predicted correctly. Consequently, more false predictions occur during the interictal epochs. The tight dependency between sensitivity and the false prediction rate holds for every prediction method.

Three shortcomings exist which need to be resolved for further development of seizure prediction methods:

1 Different assessment criteria of seizure prediction methods exist and the term “seizure prediction” is not unequivocally defined.

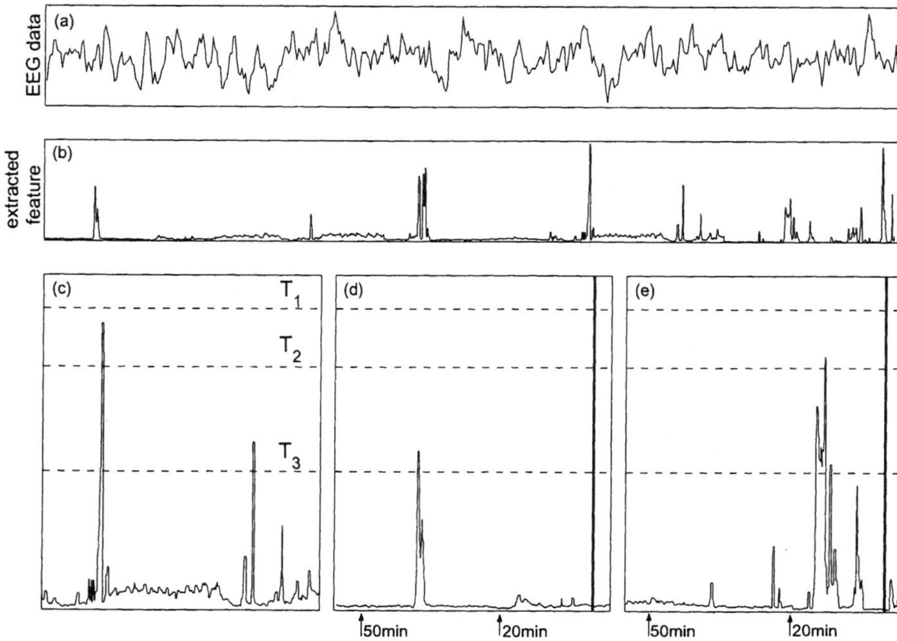

Figure 5.1. Dependency between sensitivity and false prediction rate. The upper and middle panels display an example for EEG data (a) and an extracted feature (b) used by the seizure prediction method "increments of the accumulated energy". Below, one hour interictal (c) and two hours pre-ictal epochs are shown (d,e). Bold vertical lines mark seizure onsets. Upward crossing of a threshold (dashed line) triggers an alarm. Three different thresholds illustrate the dependency between sensitivity and false prediction rate: For T_1 no alarm occurs neither during pre-ictal nor interictal epochs, meaning zero sensitivity and zero false predictions. Threshold T_2 leads to the correct prediction of the second seizure in (e) in a time interval 20 minutes before seizure onset, at the expense of one false prediction during the interictal epoch in (c). Decreasing the threshold to T_3 in order to predict the first seizure in (d), produces another false alarm. Evaluation of a prediction method should require the simultaneous assessment of both sensitivity and false prediction rate.

2 Generally little attention has been paid to the dependency between sensitivity and false prediction rate. The performances of most prediction methods summarized in table 5.1 for example are characterized only by their sensitivity without calculation of the false prediction rate.

3 All prediction methods were developed and tested on different EEG data pools, making it difficult to compare their performance.

Osorio et al. suggested in 1998 that prediction methods should be evaluated by both, sensitivity *and* false prediction rate ([Osorio et al., 1998]). We extended this approach and developed an assessment criterion called "seizure prediction characteristic" ([Winterhalder et al., 2003a]). It takes into account

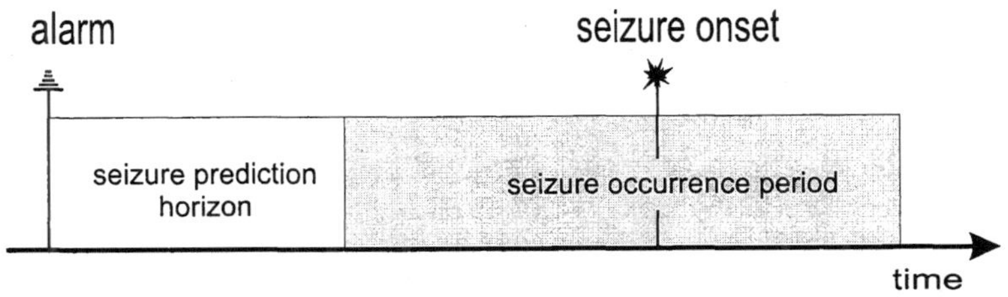

Figure 5.2. Definition of a correct prediction: The seizure doesn't occur before the end of the seizure prediction horizon (SPH). This time interval is followed by the seizure occurrence period (SOP), during which the seizure occurs, but the exact point of time is unknown. Seizures outside of any SOP are not predicted and therefore are considered false negatives. Alarm signals without a seizure in the following seizure occurrence period are false predictions.

statistical and clinical considerations and makes it possible to assess and compare different seizure prediction methods.

In the following, we present a suitable terminology for seizure prediction and the motivation, definition and eventually the calculation of the seizure prediction characteristic. An application of the assessment criterion on three seizure prediction methods can be found in this volume ([Winterhalder et al., 2003b]).

2. The Seizure Prediction Characteristic

2.1. Terminology

A seizure prediction method has to forecast an impending epileptic seizure by raising an alarm in advance of seizure onset. A perfect prediction method would indicate the exact point in time when a seizure is to occur. This ideal behavior is not expected for current prediction methods analyzing EEG data. The uncertainty can be considered by use of the seizure occurrence period SOP, which is defined as a time period during which the seizure is to be expected (Fig. 5.2). In addition, in order to render a therapeutic intervention possible, a minimum window of time between the alarm raised by the prediction method and the beginning of SOP is essential. This time window is denoted as seizure prediction horizon SPH. Taking into account the two time periods SPH and SOP, a correct prediction is defined as follows: After the alarm signal, during SPH, no seizure has occurred yet. During SOP a seizure occurs. The exact time of seizure onset may vary within SOP, thereby reflecting the uncertainty of the prediction. Seizures outside of any SOP are not predicted by the system and therefore classified as false negatives. Alarm signals without a seizure during SOP are false predictions.

Two measures to describe a prediction method performance for given SPH and SOP include:

- sensitivity, defined as the fraction of correctly predicted seizures of the total seizures
- false prediction rate, the number of false predictions per time interval

As discussed above, these measures are not independent.

2.2. Clinical Considerations

Single false predictions are not avoidable in a realistic setting. Measurements in large complex systems like the human brain are subject to fluctuations likely producing false alarms, if the investigated time interval is long enough. Should false alarms occur, the patient prepares for the arising seizure in vain. In the case of electrical stimulations or administration of drugs, unnecessary side effects may occur. For a too large number of false predictions per time interval, patients will not take further alarms seriously or will suffer from psychological stresses. Side effects of repeated interventions will add up and may lead to a neurophysiological impairment. Depending on the patient and chosen intervention system, a maximum false prediction rate FPR_{max} has to be defined that is acceptable from a clinical point of view.

The average seizure incidence may be a basis through which reasonable values for FPR_{max} can be chosen. [Bauer and Burr, 2001] evaluated seizure diaries of 63 patients resistant to anticonvulsant treatment. Based on nearly nine years of documentation and about 313 seizures per patient on average, the mean seizures rate amounts to 3 per month. Reduction of anti-epileptic drugs, e.g., during presurgical monitoring, leads to increased seizure frequencies. [Haut et al., 2002] investigated seizure clustering for 91 patients with medically intractable epilepsy who underwent monitoring for presurgical evaluation. The averaged maximal number of seizures in a 24-hour period during monitoring increased to 3.6 seizures per day compared to the low number under normal conditions. Higher values of FPR_{max} are questionable with respect to possible clinical applications. Even if all seizures can be predicted correctly, at least 50% of all alarms would be false alarms for patients during monitoring. This percentage increases to 97% for epileptic patients under normal conditions.

Similar constraints exist for SPH and SOP, depending on patient and intervention system. The application of anticonvulsive drugs needs a certain time period until taking effects due to the distribution of the drug, passage through the blood brain barrier and effect on the target neuron. Here, a minimum seizure prediction horizon SPH_{min} is required. Electrical stimulation is supposed to be fast acting and may require only a few seconds. If the patient is only warned, SPH_{min} increases to tens of seconds - enough time for the patient to leave a dangerous situation.

Because the exact point of time for seizure onset is unknown, interventions should have effects lasting for the whole seizure occurrence period. Too large durations for SOP may require further administrations of anticonvulsive drugs or long electrical stimulations. In the case of a warning system, the patients' psychological stress increases with longer SOP, because a seizure is expected at any moment during this time interval. Thus, too long SOP would increase the patient's anxiety. The physiological and psychological stress determines an upper bound for SOP, the maximum seizure occurrence period SOP_{max}.

2.3. Statistical Considerations

To be regarded as a prediction method, the performance of a seizure prediction method has to be superior to a prediction in a random, periodical or another nonspecific manner, independent of any prior information. Figure 5.3 displays how seizures can be predicted correctly by chance: In general, the parameters of a seizure prediction method will be adjusted to increase sensitivity until the false prediction rate equals the upper bound FPR_{max}. Then, during a small interictal time interval I the probability for an alarm is $p = FPR_{max}\, I$. Observing a longer time interval W the probability for at least one alarm can be calculated as follows:

$$\begin{aligned} p(\text{no alarm in } I) &= 1 - FPR_{max}\, I \\ p(\text{no alarm in } W) &= (1 - FPR_{max}\, I)^{\frac{W}{I}} \\ p(\text{at least one alarm in } W) &= 1 - (1 - FPR_{max}\, I)^{\frac{W}{I}} \\ p(\text{at least one alarm in } W \gg I) &\approx 1 - e^{-FPR_{max}\, W} \end{aligned}$$

With $W = SOP$ this is exactly the sensitivity S of a random prediction method, because it is the probability of at least one alarm during the seizure occurrence period.

A periodical prediction method raises alarms regularly after a certain period of time. If during interictal phases the false prediction rate equals FPR_{max}, the probability and therefore sensitivity S for an alarm during the seizure occurrence period SOP is

$$S = \min\{FPR_{max}\, SOP\, ,\, 100\%\}$$

For large values of SOP or FPR_{max}, both the random and the periodical prediction method achieve high sensitivities approaching 100% (Fig. 5.3). This happens independently of the value for the seizure prediction horizon. Regarding a maximum false prediction rate of 1.0 false predictions per hour (FP/h) and a seizure occurrence period of 50 minutes, the random prediction method achieves a sensitivity of 57% and the periodical prediction method a sensitivity of 83%. Hence, for too high maximum false prediction rates or too long seizure occurrence periods, the performance of any specific seizure prediction method can not be distinguished from the results of these unspecific prediction methods.

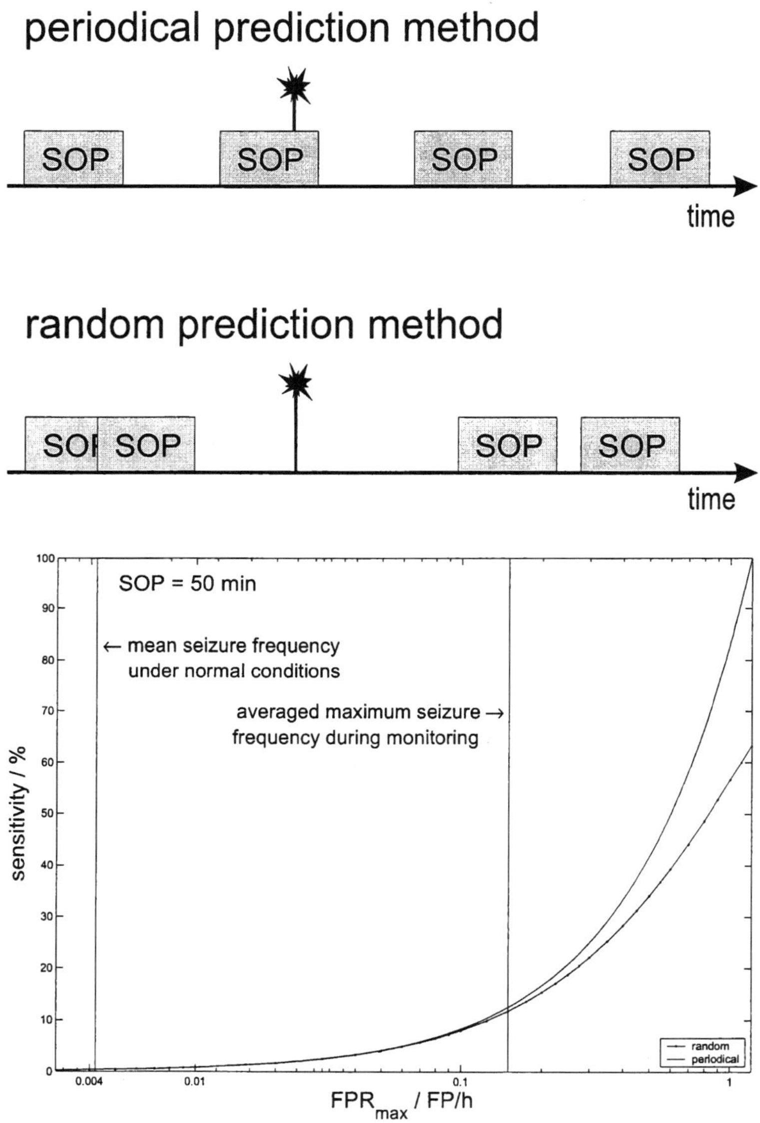

Figure 5.3. Unspecific prediction methods. Upper panel: Periodical prediction method raising alarms after a certain period of time. The random prediction method raises alarms by chance. Since seizure occurrence periods can have an overlap, sensitivity is a bit worse than for the periodical prediction method. Lower panel: Sensitivity depending on FPR_{max} for $SOP = 50$ minutes. Both methods converge to high sensitivity values for too high FPR_{max}. For example with $FPR_{max} = 1$ FP/h, sensitivity amounts to 83% for the periodical prediction method.

2.4. Assessment Criterion: The Seizure Prediction Characteristic

The values for FPR_{max}, SPH_{min}, and SOP_{max} depend on a particular clinical application i.e. a patient and an intervention system. This is generally unknown during the development of a seizure prediction method. Therefore, the method's sensitivity should not be calculated for a fixed setting but for a reasonable range of values for FPR_{max}, SPH, and SOP, leading to the seizure prediction characteristic

$$S = S(FPR_{max}, SPH, SOP).$$

This approach enables the assessment and comparison of seizure prediction methods independently of any particular clinical application. As a minimum requirement, a prediction method should be superior to unspecific methods like the random or periodical one by achieving a significant higher seizure prediction characteristic.

The calculation of the seizure prediction characteristic to evaluate a prediction method comprises five steps:

1 Specification of the number of maximum tolerated false predictions during interictal periods FPR_{max}, SPH, and SOP.

2 Adjustment of parameters of the prediction method, for example the value of a threshold, until the false prediction rate equals FPR_{max} for every single patient. Interictal data sets of at least $1/FPR_{max}$ duration for each patient are required for this procedure.

3 Calculation of sensitivity S using the pre-ictal data sets of each patient.

4 Averaging the values of sensitivity for all patients.

5 Repetition of these steps for a reasonable range of values for FPR_{max}, SPH, and SOP. Eventually the seizure prediction characteristic $S(FPR_{max}, SPH, SOP)$ can be estimated.

3. Conclusion

We suggest application of the seizure prediction characteristic as a function of sensitivity S and the maximum false prediction rate FPR_{max}, the seizure prediction horizon SPH, and seizure occurrence period SOP, to determine the performance of a seizure prediction method. In this way, it is possible to assess and to compare prediction methods and to choose a suitable method for a particular patient and type of intervention. The minimum requirement for a seizure prediction method is a significantly higher seizure prediction characteristic than for unspecific prediction methods.

References

[Bauer and Burr, 2001] Bauer, J. and Burr, W. (2001). Course of chronic focal epilepsy resistant to anticonvulsant treatment. *Seizure*, 10:239–46.

[De Clercq et al., 2002] De Clercq, W., Lemmerling, P., Van Huffel, S., and Van Paesschen, W. (2002). Evaluation of methods for determining a pre-ictal state in scalp-eeg measurements. *Biosignal Processing*, 10.

[Haut et al., 2002] Haut, S. R., Swick, C., Freeman, K., and Spencer, S. (2002). Seizure clustering during epilepsy monitoring. *Epilepsia*, 43(7):711–715.

[Iasemidis et al., 2001] Iasemidis, L., Pardalos, P., Sackellares, J., and Shiau, D. (2001). Quadratic binary programming and dynamical system approach to determine the predictibilty of epileptic seizures. *Journal of Combinatorial Optimization*, 5:9–26.

[Jerger et al., 2001] Jerger, K. K., Netoff, T. I., Francis, J. T., Sauer, T., Pecora, L., Weinstein, S. L., and Schiff, S. J. (2001). Early seizure detection. *Journal of Clin. Neurophysiology*, 18:259–268.

[Le Van Quyen et al., 2000] Le Van Quyen, M., Adam, C., Martinerie, J., Baulac, M., Clemenceau, S., and Varela, F. (2000). Spatio-temporal characterizations of non-linear changes in intracranial activities prior to human temporal lobe seizures. *European Journal of Neuroscience*, 12:2124–2134.

[Le Van Quyen et al., 1999] Le Van Quyen, M., Martinerie, J., Baulac, M., and Varela, F. (1999). Anticipating epileptic seizures in real time by a non-linear analysis of similarity between EEG recordings. *Neuroreport*, 10:2149–2155.

[Le Van Quyen et al., 2001] Le Van Quyen, M., Martinerie, J., Navarro, V., Boon, P., D'Have, M., Adam, C., Renault, B., Varela, F., and Baulac, M. (2001). Anticipation of epileptic seizures from standard EEG recordings. *The Lancet*, 357:183–188.

[Lehnertz et al., 2001] Lehnertz, K., Andrzejak, R., Arnhold, J., Kreuz, T., Mormann, F., Rieke, C., Widman, G., and Elger, C. (2001). Nonlinear EEG analysis in epilepsy: Its possible use for interictal focus localization, seizure anticipation, and prevention. *J. Clin. Neurophysiol.*, 18:209–222.

[Lehnertz and Elger, 1998] Lehnertz, K. and Elger, C. (1998). Can epileptic seizures be predicted? Evidence from nonlinear time series analysis of brain electrical activity. *Phys. Rev. Lett.*, 80:5019–5022.

[Litt and Echauz, 2002] Litt, B. and Echauz, J. (2002). Prediction of epileptic seizures. *The Lancet Neurology*, 1:22–30.

[Litt and Lehnertz, 2002] Litt, B. and Lehnertz, K. (2002). Seizure prediction and the preseizure period. *Current Opinion in Neurology*, 15:173–177.

[Martinerie et al., 1998] Martinerie, J., Adam, C., Quyen, M. L. V., Baulac, M., Clemenceau, S., Renault, B., and Varela, F. (1998). Epileptic seizures can be anticipated by non-linear analysis. *Nature Medicine*, 4:1173–1176.

[Mormann et al., 2003] Mormann, F., Andrzejak, R., Kreuz, T., Rieke, C., David, P., Elger, C., and Lehnertz, K. (2003). Automated detection of a pre-seizure state based on a decrease in synchronization in intracranial EEG recordings from epilepsy patient. *Phys. Rev. E*,67:021912.

[Navarro et al., 2002] Navarro, V., Martinerie, J., Le Van Quyen, M., Clemenceau, S., Adam, C., Baulac, M., and Varela, F. (2002). Seizure anticipation in human neocortical partial epilepsy. *Brain*, 125:640–655.

[Osorio et al., 1998] Osorio, I., Frei, M., and Wilkinson, S. (1998). Real-time automated detection and quantitative analysis of seizures and short-term prediction of clinical onset. *Epilepsia*, 39(6):615–27.

[Schindler et al., 2002] Schindler, K., Wiest, R., Kollar, M., and Donati, F. (2002). EEG analysis with simulated neuronal cell models helps to detect pre-seizure changes. *Clin. Neurophysiol.*, 113:604–614.

[Winterhalder et al., 2003a] Winterhalder, M., Maiwald, T., Voss, H., Aschenbrenner-Scheibe, R., Timmer, J., and Schulze-Bonhage, A. (2003a). The seizure prediction characteristic - a general framework to assess and compare seizure prediction methods. *Epilepsy and Behavior*, 4(3):318–325.

[Winterhalder et al., 2003b] Winterhalder, M., Maiwald, T., Voss, H. U., Aschenbrenner-Scheibe, R., Schulze-Bonhage, A., and Timmer, J. (2003b). Seizure prediction methods - assessment and comparison by means of the seizure prediction characteristic. In this volume. Kluwer Academic Publishers.

Chapter 6

SEIZURE PREDICTION METHODS

Assessment and comparison of three methods by means of the seizure prediction characteristic

M. Winterhalder
FDM, Center for Data Analysis and Modeling, University of Freiburg, Eckerstr. 1, 79104 Freiburg, Germany
winterhm@fdm.uni-freiburg.de

T. Maiwald
FDM, Center for Data Analysis and Modeling, University of Freiburg, Eckerstr. 1, 79104 Freiburg, Germany
maiwald@fdm.uni-freiburg.de

H.U. Voss
FDM, Center for Data Analysis and Modeling, University of Freiburg, Eckerstr. 1, 79104 Freiburg, Germany
hv@fdm.uni-freiburg.de

R. Aschenbrenner-Scheibe
Epilepsy Center, University of Freiburg, Breisacher Str. 64, 79106 Freiburg, Germany
aschenbs@nz.ukl.uni-freiburg.de

A. Schulze-Bonhage
Epilepsy Center, University of Freiburg, Breisacher Str. 64, 79106 Freiburg, Germany
schulzeb@nz.ukl.uni-freiburg.de

J. Timmer
FDM, Center for Data Analysis and Modeling, University of Freiburg, Eckerstr. 1, 79104 Freiburg, Germany
jeti@fdm.uni-freiburg.de

Abstract Several methods have been suggested to predict the onset of epileptic seizures from EEG data. We evaluated the performance of three predictions methods: the "dynamical similarity index", the "effective correlation dimension" and an extended, prospective version of the "accumulated energy". These prediction methods were applied on a large pool of intracranial EEG data from 21 patients. Altogether, 582 hours EEG data and 88 seizures were investigated.

The "seizure prediction characteristic" was used as assessment criterion. It considers the strong dependency between sensitivity and the false prediction rate. For a rate of 1 to 3.6 false predictions per day, the similarity index yields a sensitivity between 21% and 42%, which was the best result of the three examined prediction methods. The extended version of the accumulated energy achieves a sensitivity between 18% and 31%, the effective correlation dimension between 13% and 30%.

Keywords: Epilepsy, seizure prediction, seizure prediction characteristic, intracranial EEG data, false prediction rate

1. Introduction

The major characterizing symptoms of epilepsy are seizures affecting control of movement, consciousness or perception. The unforeseeable occurrence of seizures leads to great mental strain for the patients. About 0.7% of the population in the industrial countries suffer from this disease [Hauser et al., 1993].

It is possible to control epileptic seizures using anticonvulsive drugs in 70% of all patients. In a subgroup of 18% of all patients, seizure control can be achieved by epilepsy surgery [Wiebe et al., 2001]. However, 12% of all patients can neither be treated by medication nor by epilepsy surgery.

Therefore, new therapeutic options are necessary. One option to treat epilepsy might be the development of a device to prevent a seizure in advance of seizure onset, called "brain defibrillator" in analogy to cardiac defibrillators [Milton and Jung, 2003]. For this purpose, two important problems have to be solved. One problem is how to control the generation of epileptic activity. This might be solved by using interventions like electrical stimulation or drug delivery using a minipump [Nicolelis, 2001].

A different challenge is to predict the onset of an upcoming seizure. Here, special interest is focussed on the prediction of epileptic seizures from intracranial or scalp EEG data by methods based on nonlinear and linear time series analysis. Several investigations claim promising results [Lehnertz and Elger, 1995, Lehnertz and Elger, 1998, Lehnertz et al., 2001, Iasemidis et al., 1990, Iasemidis et al., 2001, Le van Quyen et al., 1999, Le van Quyen et al., 2001b, Le van Quyen et al., 2001a, Navarro et al., 2002, Litt et al., 2001, Mormann et al., 2000, Schindler et al., 2002, Jerger et al., 2001]. However, most of these studies have been evaluated using only a few seizures, a low number of patients or insufficient interictal, seizure-free data sets. Furthermore, no recognized performance standards to assess prediction methods exist [Litt and Lehnertz, 2002], making it difficult to compare different methods.

We present the assessment and comparison of three univariate seizure prediction methods: the "dynamical similarity index" [Le van Quyen et al., 1999, Le van Quyen et al., 2001b, Le van Quyen et al., 2001a, Navarro et al., 2002], the "effective correlation dimension" [Lehnertz and Elger, 1995, Lehnertz and Elger, 1998, Lehnertz et al., 2001] and an extended, prospective version of the "accumulated energy" [Litt et al., 2001].

The "seizure prediction characteristic" was used as assessment criterion [Winterhalder et al., 2003, Maiwald et al., 2003]. It mirrors the functional relation between sensitivity S and three measures characterizing a prediction method: First, the maximum permitted number of false predictions per hour FPR_{max} during interictal states. Second, the seizure prediction horizon SPH, representing the time between an alarm and the earliest possible seizure onset. And third, the seizure occurrence period SOP during which a seizure is supposed to occur at any time.

We calculated the seizure prediction characteristic for the three prediction methods using intracranial EEG data of 21 patients. Our data pool consist of 2-5 seizures and at least 24 hours interictal EEG data for each patient, altogether 88 seizures and 582 hours EEG data. It is intended to publish this still growing data pool as an open source for the development and evaluation of prediction methods.

The EEG data pool is described more detailed in the next section. Part 3 summarizes the concepts of the three investigated prediction methods. The steps to calculate the seizure prediction characteristic are reported in part 4, the results are given in part 5.

2. EEG data and patients characteristics

The prediction methods were applied using intracranial EEG data from 21 patients suffering from pharmacorefractory focal epilepsy of temporal and extratemporal origin. These data were recorded during presurgical epilepsy monitoring using a sampling rate of 256 Hz or 512 Hz.

Invasive electrodes were used in order to study the EEG data at a high signal to noise ratio. Depth electrodes were implanted stereotactically and subdural electrodes via burr holes or open craniotomy. Preparation of the data included bandpass filtering in the frequency domain between 0.5 and 120 Hz respectively 80 Hz for the effective correlation dimension. A 50 Hz notch filter eliminated possible line noise.

A certified epileptologist selected six electrodes of all implanted electrodes, which were referenced to an electrode displaying a minimal amount of epileptic activity. For every patient, 2-5 seizures (mean 4.2) were examined, including pre-seizure EEG data of at least 50 min duration. This pre-seizure data sets were used to calculate the sensitivity of the prediction methods.

To evaluate false prediction rates, long-term interictal, seizure-free EEG data are necessary. We investigated 24 hours interictal data for each patient. For 13 patients, 24 hours of contiguous interictal recordings were available. In the remaining cases, seizure-free periods were shorter than 24 hours, thus a few contiguous intervals were combined to at least 24 hours duration.

3. Seizure Prediction Methods

We examined the performance of three predictions methods: the dynamical similarity index, the effective correlation dimension and an extended, prospective version of the accumulated energy. The main steps to calculate these prediction methods are summarized in this section.

3.1. Dynamical Similarity Index

The dynamical similarity index was implemented as introduced [Le van Quyen et al., 1999]. Its capability to predict seizures several minutes in advance was demonstrated in several studies [Le van Quyen et al., 2001b, Le van Quyen et al., 2001a, Navarro et al., 2002]. The similarity index algorithm compares the dynamical behavior in a sliding window S_t to a fixed reference window S_{ref}, which is chosen far away from any seizure.

As a first step, new time series $I_n, n \in \mathbb{N}$, are constructed by computing time intervals between two positive zero-crossings of the EEG data. A delay embedding with embedding dimension m and delay τ leads to $A_n = (I_n, I_{n-\tau}, ..., I_{n-(m\tau)+1})$. Next, a singular value decomposition for the trajectory matrix $\mathbf{A}(S_{ref})$ of the reference window is applied. The trajectory matrices $\mathbf{A}(S_t)$ and $\mathbf{A}(S_{ref})$ are projected on the principal axes of the reference window, yielding $\mathbf{X}(S_t)$ respectively $\mathbf{X}(S_{ref})$. A random selection $\mathbf{Y}(S_{ref})$ of $\mathbf{X}(S_{ref})$ is compared to $\mathbf{X}(S_t)$ via the cross-correlation integral

$$C(S_{ref}, S_t) = \frac{1}{N_{ref}N_t} \sum_{i=1}^{N_{ref}} \sum_{j=1}^{N_t} \Theta\left(r - ||Y_i(S_{ref}) - X_j(S_t)||\right).$$

Here, Θ denotes the Heaviside step function, $||\cdot||$ the euclidian norm, and N_{ref} respectively N_t the number of vectors of the reference and sliding window. The distance r is chosen as the 30% quantile of the cumulative neighborhood distribution of the reference window. Finally, the dynamical similarity index $\gamma(S_t)$ is given by

$$\gamma(S_t) = \frac{C(S_{ref}, S_t)}{\sqrt{C(S_{ref}, S_{ref})C(S_t, S_t)}}.$$

Threshold crossing with the constraint of a minimum crossing time of 150 seconds was used as alarm signal. The threshold was varied.

3.2. Effective Correlation Dimension

Based on the correlation dimension D_2, which is an estimator for the fractal dimension of the attractor of a deterministic dynamical system, the performance of the effective correlation dimension D_2^{eff} to predict seizures was investigated [Lehnertz and Elger, 1995, Lehnertz and Elger, 1998, Lehnertz et al., 2001].

The calculation starts with a delay embedding of the EEG time series for several dimension values up to $m = 25$, leading to $\vec{x}_m(t)$. The correlation sum

$$C_m(r) = \frac{1}{N(N-1)} \sum_{i \neq j} \Theta(r - ||\vec{x}_m(i) - \vec{x}_m(j)||)$$

is calculated for a range of the radius r. Here, $||\cdot||$ denotes the maximum norm. The correlation dimension is defined as

$$D_2 = \lim_{r \to 0} \frac{\mathrm{d}\log C_m(r)}{\mathrm{d}\log(r)}.$$

The limit requires a proper scaling region, which is not necessarily given for measured data. To overcome this problem, the authors introduced an operational method leading to the so called effective correlation dimension D_2^{eff}. This measure is applied on the EEG data using a sliding window technique.

In [Lehnertz et al., 2001] so called "dimension drops" were evaluated, characterized by the time interval t_{drop} and maximal deviation d_{drop}, the values of D_2^{eff} dropped under a threshold. The average of D_2^{eff} during interictal periods was used as threshold. A dimension drop was defined as being predictive if it directly precedes a seizure onset and the dropping parameters extend the maximum dropping parameters during interictal periods. Use of these criteria leads to no false predictions, but in our investigation only one of all 88 seizures was preceded by such a predictive dimension drops. The long interictal EEG data used in our study might be a possible explanation for this result. Allowing false predictions increased sensitivity and was necessary to calculate the seizure prediction characteristic. This was achieved by varying the dropping parameters t_{drop} and d_{drop}.

3.3. Increments of Accumulated Energy

Recently, the discriminating power of the accumulated energy with respect to pre-ictal and interictal periods of 50 min duration was investigated [Litt et al., 2001]. About 90% of the pre-ictal and 88% of the interictal periods were classified correctly. Unfortunately, this method requires knowledge of seizure onset which is not given in a prospective analysis. Based on these promising results, we examined an extended, prospective version, the so called "increments of the accumulated energy."

The accumulated energy $AE(k)$ is based on the "average energy"

$$E_k = \text{mean}(x_i^2) \text{ for time window } k \ (k = 1, 2, ...)$$

calculated for a time window of 1.25 s, x_i denotes the electrode potential of sample i. Two consecutive time windows are shifted by 0.45 s. The accumulated energy respectively the increments of the accumulated energy iAE are defined as

$$AE_m = \frac{1}{10} \sum_{k=10m-9}^{10m} E_k + AE_{m-1},$$

$$iAE_m = \frac{1}{10} \sum_{k=10m-9}^{10m} E_k = AE_m - AE_{m-1}.$$

A higher slope of AE corresponds to higher increments iAE. Using a median filter over 90 seconds ensures that only permanent changes of these increments lead to different values of iAE. Threshold crossing of iAE was used as alarm signal and the threshold value was varied.

4. Calculation of the seizure prediction characteristic

The seizure prediction characteristic [Winterhalder et al., 2003] was used to assess the three prediction methods. It is a means to evaluate and compare seizure prediction methods. It is based on the functional relation of sensitivity S on three measures characterizing the performance of a given seizure prediction method, namely the maximum false prediction rate FPR_{max}, the seizure occurrence period SOP and the seizure prediction horizon SPH. The theoretical background is also reported in another paper in this volume [Maiwald et al., 2003].

The calculation of the seizure prediction characteristic for a given prediction method consists of five steps:

1 Specify the number of maximum tolerated false predictions during interictal periods, the maximum false prediction rate FPR_{max}. Fix the seizure occurrence period SOP, which is the time during which a seizure is supposed to occur and the seizure prediction horizon SPH, which mirrors the time between an alarm and the earliest possible seizure onset.

2 Adjust parameters of the prediction method, e.g. the value of a threshold or the time interval, a feature has to stay below a threshold until the maximum false prediction rate is reached during interictal periods, individually for each patient. Interictal data sets of at least $1/FPR_{max}$ duration are required for this procedure.

3 Calculate the sensitivity using the pre-ictal data sets of each patient.

4 Average the values of sensitivity for all patients.

5 Finally, repeat these steps for a reasonable range of values for the maximum false prediction rate FPR_{max}, the seizure occurrence period SOP and prediction horizon SPH.

This procedure leads to the seizure prediction characteristics given by sensitivity S depending on FPR_{max}, SOP and SPH,

$$S(FPR_{max}, SOP, SPH).$$

5. Results and Discussion

As the seizure prediction characteristic depends on three different measures, it is necessary to fix at least one of them to display the result. In the following, the assessment and comparison of the three prediction methods is presented by the seizure prediction characteristic depending on, first the maximum false prediction rate, second the seizure occurrence period, and third the seizure prediction horizon. The values of sensitivity S of all three seizure prediction methods are compared to each other and to two unspecific prediction methods, the periodical and the random prediction methods [Maiwald et al., 2003]. Finally, dependence of sensitivity S on two measures (FPR_{max} and SOP) is shown.

5.1. Sensitivity depending on FPR$_{\text{max}}$

First, dependence of sensitivity S on the maximum false prediction rate FPR_{max} for the dynamical similarity index ($\circ$), the increments of accumulated energy ($\diamond$) and the effective correlation dimension ($\times$) is shown (Fig. 6.1).

The seizure occurrence period was fixed to 30 min. and the seizure prediction horizon to five seconds corresponding to a very fast intervention system. The

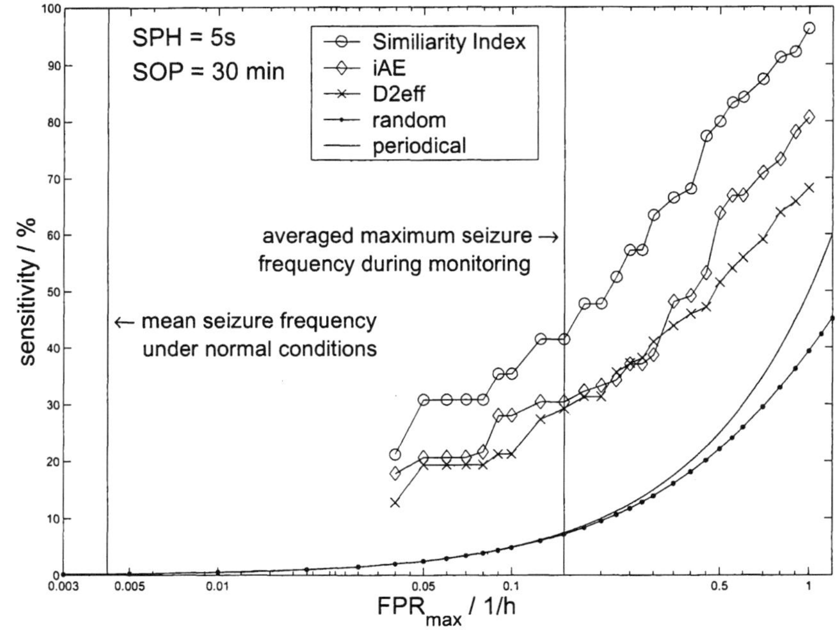

Figure 6.1. Dependence of sensitivity on the maximum false prediction rate for given seizure occurrence period and prediction horizon, for the three examined prediction methods. The dotted line displays the performance of the random, the solid line of the periodical prediction method. All three prediction methods yield better results than the two unspecific ones. The dynamical similarity index achieves highest sensitivity values.

vertical lines mark the mean seizure frequency under normal conditions (left) [Bauer and Burr, 2001] and the averaged maximum seizure frequency during presurgical monitoring (right) [Haut et al., 2002].

The logarithmically scaled maximum false prediction rate covers three regions. Evaluation of false prediction rates corresponding to the mean seizure frequency under normal conditions of about three seizures per months requires interictal EEG data of several days. The smallest value for FPR_{max}, which can be evaluated using an EEG data pool of 24 hours interictal data for each patient, amounts to one false prediction per day respectively 0.04 false predictions per hour (FP/h).

For maximum false prediction rates smaller than the averaged maximum seizure frequency during monitoring, the dynamical similarity index achieves the best result, sensitivity amounts to 21% to 42%. The increments of accumulated energy reach sensitivity values between 18% - 31%, the effective correlation dimension between 13% - 30%.

For higher values of the maximum false prediction rate up to one false prediction per hour, sensitivity of all three prediction methods achieves high values

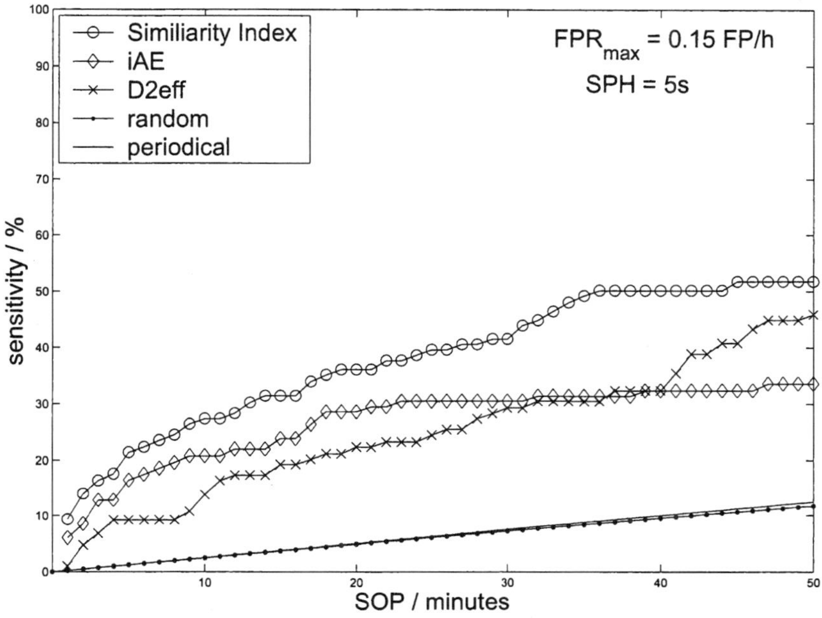

Figure 6.2. Dependence of sensitivity on the seizure occurrence period for given maximum false prediction rate and seizure prediction horizon. Again, all three prediction methods are superior to the unspecific prediction methods (dotted and solid line). The dynamical similarity index achieves the best result for all evaluated seizure occurrence periods.

of 70% - 100%. But accepting a false prediction rate of e.g. one false prediction in every five hours, 57% of the alarm events would be false ones, if the patient suffers from a seizure frequency corresponding to averaged maximum seizure frequency during monitoring. Furthermore, compared to the mean seizure frequency under normal conditions, 98% of the alarm events would be incorrect. Hence, maximum false prediction rates higher than the averaged maximum seizure frequency during monitoring are questionable.

All three prediction methods yield better results than the unspecific prediction methods independent of the investigated maximum false prediction rate.

5.2. Sensitivity depending on *SOP*

Next, values of sensitivity S depending on the seizure occurrence period are given. The maximum false prediction rate is fixed to 0.15 FP/h and the seizure prediction horizon again to 5 seconds (Fig. 6.2). The dynamical similarity index again yields the best result. For small seizure occurrence periods, the increments of accumulated energy are superior to the effective correlation dimension.

For seizure occurrence periods longer than 36 min, sensitivity of the dynamical similarity index increases slower than the unspecific prediction methods.

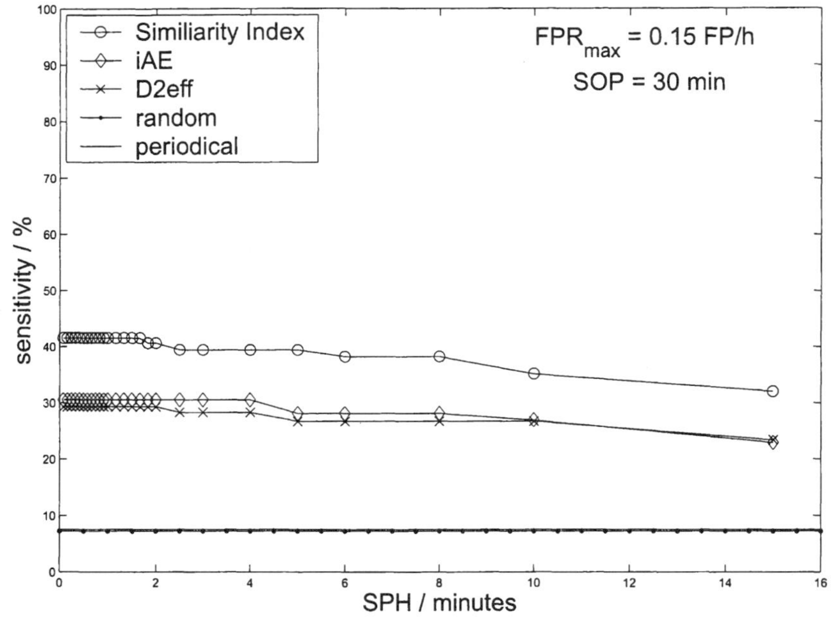

Figure 6.3. Sensitivity depending on the seizure prediction horizon for fixed values of the maximum false prediction rate and seizure occurrence period. As the random and periodical prediction methods do not depend on the prediction horizon, their values for sensitivity are constant (dotted and solid line).

Thus, the increase in sensitivity for larger values of SOP is rather a statistical property than a performance feature of the prediction method. This effect can also be detected for the increments of the accumulated energy for $SOP > 20$ min.

5.3. Sensitivity depending on SPH

The next example shows dependence of sensitivity S on the seizure prediction horizon SPH (Fig. 6.3). The seizure prediction horizon corresponds to the actual prediction performance of a prediction method. The maximum false prediction rate is again fixed to 0.15 false predictions per hour, the seizure occurrence period to 30 min.

All three prediction methods show constant sensitivity values for seizure prediction horizons $SPH < 2$ min. This is a promising result as this prediction time might be sufficient for most interventions, e.g. electrical stimulation.

5.4. Sensitivity depending on FPR_{max} and SOP

As a last example, sensitivity S is displayed depending on two of the three measures, the maximum false prediction rate and the seizure occurrence period.

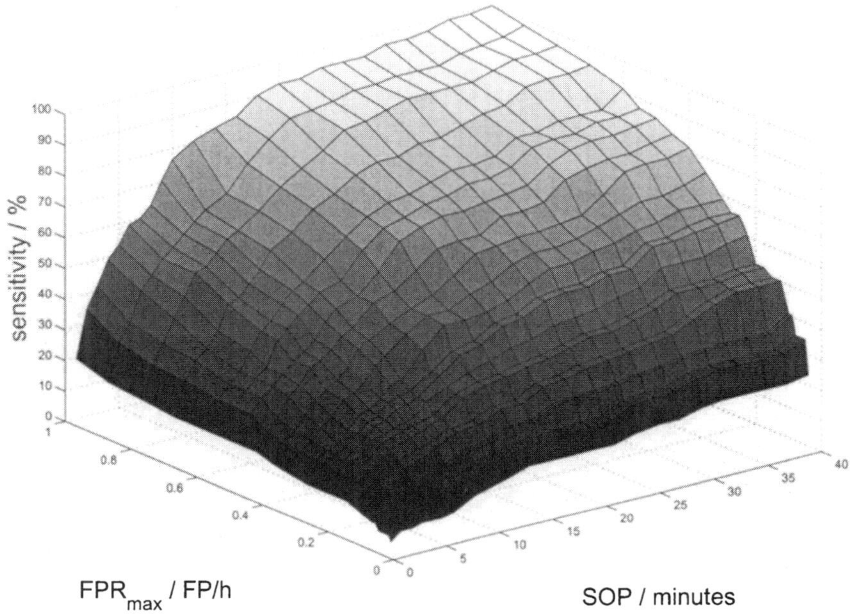

Figure 6.4. Sensitivity for the dynamical similarity index depending on the maximum false prediction rate FPR_{max} and the seizure occurrence period SOP, for a given seizure prediction horizon of 5 seconds. For several combinations of FPR_{max} and SOP, sensitivity reaches high values. But the applicability in a therapeutical device using these combinations is questionable due to the high values of the maximum false prediction rate respectively seizure occurrence period.

The seizure prediction horizon was fixed to 5 seconds. The example shows values for sensitivity for the dynamical similarity index (Fig. 6.4).

It illustrates an important fact: it is possible to achieve values for sensitivity of 80% - 100% for a wide range of the maximum false prediction rate and the seizure occurrence period. This is also shown in a gray-scaled figure for the same example (Fig. 6.5). The upper right corner corresponds to high sensitivity values. For example, for a seizure occurrence period of 40 min and one false prediction every two hours, sensitivity amounts 86%. But for a patient under normal conditions suffering from three seizures per month, 98% of the alarm events would be false ones.

The important question is, whether these combinations of values for the maximum false prediction rate, seizure occurrence period, and prediction horizon are suitable for a therapeutical device to control seizures.

6. Conclusion

We have shown the assessment and comparison of three prediction methods. This was possible by using the seizure prediction characteristic, which relates

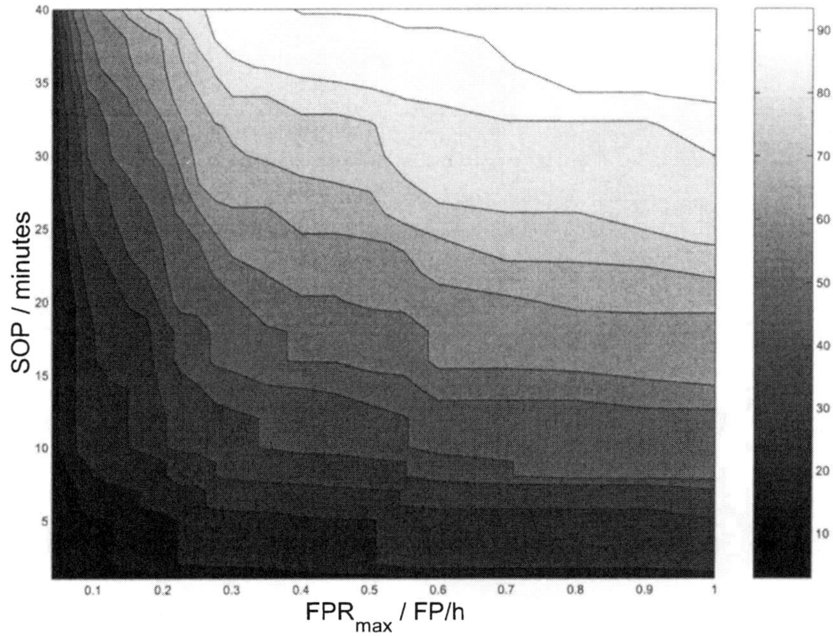

Figure 6.5. Gray-scaled plot of sensitivity for the dynamical similarity index depending on the maximum false prediction rate FPR_{max} and the seizure occurrence period SOP, for given seizure prediction horizon.

sensitivity of a prediction method to three measures characterizing important features of any prediction method.

Relating the maximum false prediction rate to the averaged maximum seizure frequency during epilepsy monitoring, only maximum false prediction rates lower than 0.15 false predictions per hour respectively three seizures per day should be considered. Even if a seizure prediction method achieves a sensitivity of 100%, at least 50% of all alarms would be false ones during presurgical monitoring. Even worse, patients under normal conditions with three seizures per month would endure 97% false predictions. Thus, higher values of the maximum false prediction rate are questionable with respect to an application in a therapeutical device controlling a seizure.

Combining these values of FPR_{max} with a seizure occurrence period of half an hour and a very short prediction horizon of 5 seconds, sensitivity of the dynamical similarity index yields values between 21% and 42%, which was the best result of the three examined prediction methods. The extended version of the accumulated energy achieves a sensitivity between 18% and 31%, the effective correlation dimension between 13% and 30%.

The results of the examined prediction methods are significant better than the performance of the unspecific random and periodical prediction methods.

This indicates that EEG data contain specific, predictive information during pre-ictal periods. However, the resulting seizure prediction characteristics are not sufficient using it in a therapeutical device to prevent epileptic seizure.

References

[Bauer and Burr, 2001] Bauer, J. and Burr, W. (2001). Course, of chronic focal epilepsy resistant to anticonvulsant treatment. *Seizure*, 10:239–46.

[Hauser et al., 1993] Hauser, W., Annegers, J., and Kurland, L. (1993). Incidence of epilepsy and unprovoked seizures in Rochester, Minnesota: 1935-1984. *Epilepsia*, 34(3):453–68.

[Haut et al., 2002] Haut, S. R., Swick, C., Freeman, K., and Spencer, S. (2002). Seizure clustering during epilepsy monitoring. *Epilepsia*, 43(7):711–715.

[Iasemidis et al., 2001] Iasemidis, L., Pardalos, P., Sackellares, J., and Shiau, D. (2001). Quadratic binary programming and dynamical system approach to determine the predictibilty of epileptic seizures. *Journal of Combinatorial Optimization*, 5:9–26.

[Iasemidis et al., 1990] Iasemidis, L., Sackellares, J., Zaveri, H., and Williams, W. (1990). Phase space topography and the Lyapunov exponent of electrocorticograms in partial seizures. *Brain Topography*, 2:187–201.

[Jerger et al., 2001] Jerger, K. K., Netoff, T. I., Francis, J. T., Sauer, T., Pecora, L., Weinstein, S. L., and Schiff, S. J. (2001). Early seizure detection. *Journal of Clinical Neurophysiology*, 18:259–268.

[Le van Quyen et al., 1999] Le van Quyen, M., Martinerie, J., Baulac, M., and Varela, F. (1999). Anticipating epileptic seizures in real time by a non-linear analysis of similarity between EEG recordings. *Neuroreport*, 10:2149–2155.

[Le van Quyen et al., 2001a] Le van Quyen, M., Martinerie, J., Navarro, V., Baulac, M., and Varela, F. (2001a). Characterizing neurodynamic changes before seizures. *Journal of Clinical Neurophysiology*, 18:191–208.

[Le van Quyen et al., 2001b] Le van Quyen, M., Martinerie, J., Navarro, V., Boon, P., D'Have, M., Adam, C., Renault, B., Varela, F., and Baulac, M. (2001b). Anticipation of epileptic seizures from standard EEG recordings. *The Lancet*, 357:183–188.

[Lehnertz et al., 2001] Lehnertz, K., Andrzejak, R., Arnhold, J., Kreuz, T., Mormann, F., Rieke, C., Widman, G., and Elger, C. (2001). Nonlinear EEG Analysis in Epilepsy: Its Possible Use for Interictal Focus Localization, Seizure Anticipation, and Prevention. *Journal of Clinical Neurophysiology*, 18(3):209–222.

[Lehnertz and Elger, 1995] Lehnertz, K. and Elger, C. (1995). Spatio-temporal dynamics of the primary epileptogenic area in temporal lobe epilepsy characterized by neuronal complexity loss. *Electroencephalography and clinical Neurophysiology*, 95:108–117.

[Lehnertz and Elger, 1998] Lehnertz, K. and Elger, C. (1998). Can epileptic seizures be predicted ? Evidence from nonlinear time series analysis of brain electrical activity. *Physical Review Letters*, 80:5019–5022.

[Litt et al., 2001] Litt, B., Esteller, R., Echauz, J., D'Alessandro, M., Shor, R., Henry, T., Pennell, P., Epstein, C., Bakay, R., Dichter, M., and Vachtsevanos, G. (2001). Epileptic seizures may begin hours in advance of clinical onset: A report of five patients. *Neuron*, 30:51–64.

[Litt and Lehnertz, 2002] Litt, B. and Lehnertz, K. (2002). Seizure prediction and the preseizure period. *Current Opinion in Neurology*, 15:173–177.

[Maiwald et al., 2003] Maiwald, T., Winterhalder, M., Voss, H. U., Aschenbrenner-Scheibe, R., Schulze-Bonhage, A., and Timmer, J. (2003). The seizure prediction characteristic. In *this volume*. Kluwer Academic Publishers.

[Milton and Jung, 2003] Milton, J. and Jung, P. (2003). Brain defibrillators: Synopsis, problems and future directions. In Milton, J. and Jung, P., editors, *Epilepsy as a Dynamic Disease*, pages 341–352. Springer.

[Mormann et al., 2000] Mormann, F., Lehnertz, K., David, P., and Elger, C. (2000). Mean phase coherence as a measure for phase synchronization and its application to the EEG of epilepsy patients. *Physica D*, 144:358–369.

[Navarro et al., 2002] Navarro, V., Martinerie, J., Le Van Quyen, M., Clemenceau, S., Adam, C., Baulac, M., and Varela, F. (2002). Seizure anticipation in human neocortical partial epilepsy. *Brain*, 125:640–655.

[Nicolelis, 2001] Nicolelis, M. (2001). Actions from thoughts. *Nature*, 409:403–407.

[Schindler et al., 2002] Schindler, K., Wiest, R., Kollar, M., and Donati, F. (2002). EEG analysis with simulated neuronal cell models helps to detect pre-seizure changes. *Clinical Neurophysiology*, 113:604–614.

[Wiebe et al., 2001] Wiebe, S., Blume, W., Girvin, J., and Eliasziw, M. (2001). A randomized, controlled trial of surgery for temporal-lobe epilepsy. *The New England Journal of Medicine*, 345(5):311–8.

[Winterhalder et al., 2003] Winterhalder, M., Maiwald, T., Voss, H. U., Aschenbrenner-Scheibe, R., Timmer, J., and Schulze-Bonhage, A. (2003). The seizure prediction characteristic. *Epilepsy and Behavior*, 4(3):318–325.

Chapter 7

CONTROLLING NEUROLOGICAL DISEASE AT THE EDGE OF INSTABILITY

John G. Milton
Department of Neurology
University of Chicago
sp1ace@ace.bsd.uchicago.edu

Jennifer Foss
Department of Psychology
University of New Orleans
jmfoss@uno.edu

John D. Hunter
Department of Neurology
University of Chicago
jdhunter@ace.bsd.uchicago.edu

Juan Luis Cabrera
Centro de Física
IVIC Venezuela
jlc@ivic.ve

Abstract Rapid advances in technology are making the dream of treating human neurological diseases with implanted electronic devices a reality. The more such devices are able to exploit the properties of intrinsic neural control mechanisms, the more effective they will be in re-establishing control in the setting of disease. Noise and time delays are ubiquitous features of the nervous system. Three observations suggest that in order to understand control in noisy neural dynamical systems with retarded variables it will be necessary to change the focus from the identification and characterization of attractors to a study of phenomena that occur near stability boundaries (i.e., "critical phenomena"): 1) multistability has been identified in simple neural loops, the onset of epileptic seizures, and human postural sway; 2) on-off intermittency and 3) power laws arise in the nervous system. These observations support the possibility of developing strategies that treat neurological disease by the addition of appropriately designed stimuli, including noise.

1. Introduction

Everyday the dream of treating epilepsy by using implanted electronic devices comes closer to reality. The term "brain defibrillator" has been given to implantable devices that detect the occurrence of an epileptic seizure and then deliver a stimulus to abort it [Milton and Jung, 2002]. The advantage of such devices is that they would be called upon only when needed and thus would free patients from the troublesome side effects of anticonvulsant medications. Major steps toward the construction of a brain defibrillator have already been made: several techniques have been developed that enable seizure occurrence to be predicted up to 30 minutes beforehand (this Conference); recent advances in technology suggest that the construction of such a device is feasible [Hetling, 2002].

Only one question remains: how is the seizure to be aborted? Obviously, the more such devices are able to exploit the neural mechanisms involved in the generation of epileptic seizures, the more effective they will be in re-establishing neural control in the setting of a seizure.

It has long been known that the application of a brief sensory or electrical stimulus just after seizure onset can sometimes abort a seizure within 2 seconds of application [Foss and Milton, 2002, Lesser et al., 1999, Milton, 2000, Motamedi et al., 2002]. Figure 7.1 A illustrates this phenomenon: a sudden noise made 1.4s after seizure onset terminates a seizure within 1.4s [Milton, 2000]. A brief electrical stimulus stops 51 % of after-discharges in human epileptic brain within 2s; 39 % after 5s [Lesser et al., 1999].

These observations are very suggestive of an underlying multistable dynamical system [Foss and Milton, 2000, Foss and Milton, 2002, Milton, 2000, Milton and Foss, 1997]. A schematic representation of a multistable dynamical system is shown in Figure 7.1B. The multiple valleys correspond to the basins of attraction for each attractor. Ridges of varying height separate the basins. These ridges correspond to the *separatrices*, or energy barriers. Each of the basins of attraction can be assessed by appropriate initialization. A brief stimulus causes the dynamical system to switch from one basin to another; in Figure 7.1B from the attractor associated with a seizure to one that is not.

Therapeutic strategies based on the principle of producing a switch between two attractors must be performed with great care. A randomly timed stimulus has low probability of causing switches [Lechner et al., 1996]. Moreover, the

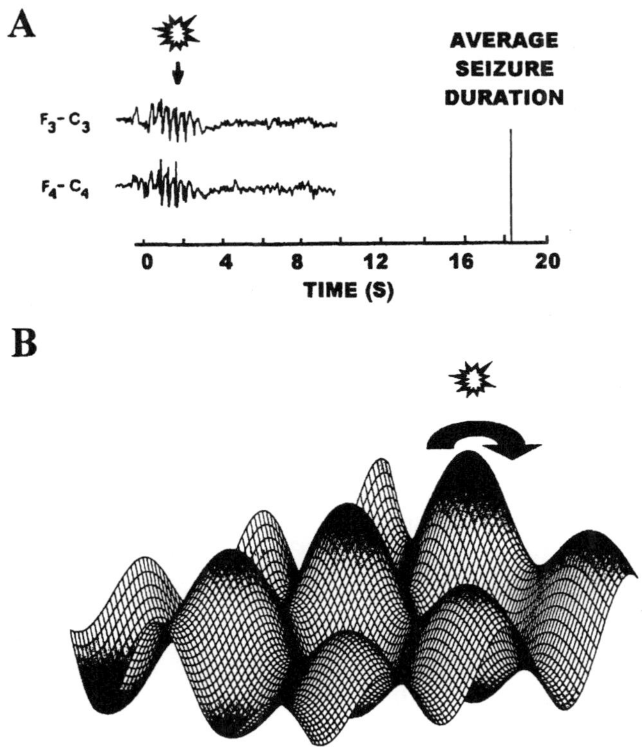

Figure 7.1. A) Stopping a seizure with an auditory stimulus [Milton, 2000]. The average seizure duration is determined for 71 spontaneously occurring seizures. The length of the auditory stimulus shortened seizure corresponds to an average of 69 seizures. B) Schematic representation of the energy landscape for a hypothetical multistable dynamical system.

use of electrical stimuli to abort a seizure is double-edged since an improperly timed electrical pulse could itself cause a seizure [Ajmone Marsan, 1972], i.e. the very event that the brain defibrillator hoped to prevent! Further complicating this approach is that random uncontrolled fluctuations ("noise") and conduction time delays are ubiquitous in the nervous system. The boundaries that separate the basins of attraction depend on the time delay and other properties of the dynamical system [Pakdaman et al., 1998]. Noise together with delays can produce dynamical behaviors that are not observed in dynamical systems that do not possess time delays, e.g. statistical periodicity [Milton and Mackey, 2000].

Here we review the current state of the art on the effects of brief stimuli and noise on the dynamics of multistable dynamical systems with retarded variables. It is important to realize that analytical tools for the study of noisy, time-delayed dynamical systems are scarce and, in general, even the evaluation of the associated probability densities is uncertain (for discussions see [Guillouzic et al., 1999, Ohira and Yamane, 2000]). Consequently we discuss these issues in the context of simple paradigms in which it has been possible to gain some insight, namely, delayed recurrent neural loops and stick balancing at the fingertip.

2. Mathematical background and outline

The fact that seizure cessation is not coincident with delivery of the aborting stimulus (Figure 7.1A) implies that memory effects are important. A variety of mechanisms can produce memory effects in neural dynamics, e.g. relative refractory periods [Milton et al., 1993], activity-dependent changes in vesicle pools [Hunter and Milton, 2001], propagation through a chain of oscillators [Ermentrout and Kopell, 1994, Kopell, 1995]). Here we focus on memory effects that arise from conduction time delays. Time delays are intrinsic components of the nervous system and arise because neurons are spatially separated and axonal conduction velocities are finite. Time delays in the central nervous system range from 10-200 ms [Gotman, 1983, Miller, 1994, Milton, 2002].

In dynamical systems that possess conduction delays, mathematical models take the form of delay differential equations (DDE). Examples include the first-order DDEs that arise in the context of neural feedback control mechanisms [an der Heiden, 1979, an der Heiden and Mackey, 1982, Mackey and Glass, 1977, Milton et al., 1989], i.e.

$$\dot{V}(t) + \alpha V(t) = f(V(t-\tau)) \tag{1}$$

where $V(t)$, $V(t-\tau)$ are, respectively, the values of the state variable (such as membrane potential) at times t, $t-\tau$. The time delay is τ, α is a rate constant, f describes the feedback, and $\dot{V}$ is the first differential of V. In order to obtain the solution of Equation 1 it is necessary to specify an initial function, ϕ, on the interval $[-\tau, 0]$. Multistability, i.e. the co-existence of multiple attractors, readily arises in dynamical systems with retarded variables [S. A. Campbell et al., 1995, Campbell et al., 1995, Foss et al., 1996, Foss et al., 1997b, Foss and Milton, 2000, Losson et al., 1993, Pakdaman et al., 1998].

Multistability is most often discussed in the context of a sub-critical Hopf bifurcation shown schematically in Figure 7.2 (other more complex mechanisms are possible, for example in bursting neurons [Izhikevich, 2000]). Sub-critical Hopf bifurcations characterize the onset of oscillations in neurons [Guttman et al., 1980] and of self-maintaining traveling waves, such as spiral waves, in model neural networks [Milton et al., 1993]. Figure 7.2 shows the stability diagram for the occurrence of a sub-critical Hopf bifurcation in a dynamical system in the presence of noise. The steady state solutions have been represented by the maxima of the stationary density, i.e., the probability density after transients have died out [Arnold, 1998, Horsthemenke and Lefever, 1984, Longtin,

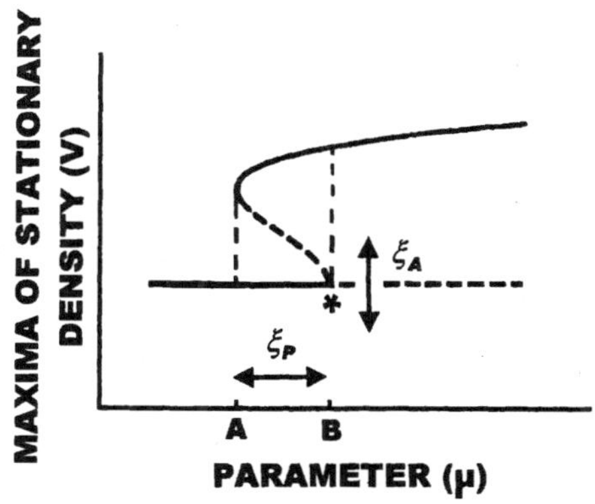

Figure 7.2. Schematic representation of a sub-critical Hopf bifurcation. The solid lines represent the stable solutions and the dashed lines the unstable solutions. This subcritical Hopf bifurcation is symmetric about the horizontal axis; we show only the upper half.

1991, Longtin et al., 1990]. In the hypothetical noise-free, or deterministic, case the probability densities are δ-functions.

For a given choice of parameter, μ, the behavior of the dynamical system shown in Figure 7.2 depends on the choice of initial condition. In the absence of a time delay the initial conditions correspond to the values of the variables, V, measured at a single point in time. However, when time delays are present the initial conditions take the form of initial functions constructed from the variables measured over an interval of time whose length equals the time delay. It is not possible to draw the space of all initial functions for a time-delayed dynamical system since the system dimension is infinite. Nonetheless we can use Figure 7.2 as a "cartoon" to provide a framework for the present discussion. When $\mu < A$, all initial conditions lead to a fixed point attractor. When $\mu > B$, all initial conditions lead to a limit cycle attractor. In the region $A \leq \mu \leq B$ the two attractors co-exist: some initial conditions lead to the fixed-point attractor, others to the limit-cycle attractor. The set of initial conditions that lead to a given attractor is the basin of attraction for that attractor. In order to be able to design perturbations to induce a switch between the attractors it is necessary to know 1) the boundaries that delineate the basins of attraction; and 2) the nature of the paths that enable a trajectory to leave one basin of attraction and enter another.

Within this framework there are two different explanations for the observation that brief stimuli can abort a seizure (Figure 7.1A: 1) changes in neural synchrony ($\mu > B$); and 2) multistability ($A \leq \mu \leq B$). Section 3 demonstrates that the latency for changes in neural synchrony due to electric stimuli and synaptic inputs is typically shorter than the latency seen in seizure abortion. Section 4 shows that the latency seen in a stimulus induced switch between attractors in a dynamical system is typically greater than τ.

Sections 5 and 6 deal with the confounding effects of neural noise. Noise is ubiquitous in the nervous system [Arielle et al., 1996, Verveen and DeFelice, 1974]. There are two ways in which noise can enter into Equation 1. Additive noise (Section 5) refers to the case in which the effect of noise is independent of the state of the system, i.e.

$$\dot{V}(t) = f(V(t), V(t-\tau), \mu) + \xi_A(t), \tag{2}$$

where $\xi_A(t)$ is, for example, Gaussian distributed white noise. In Figure 7.2, additive noise produces vertical fluctuations. Thus noise-induced transitions between the attractors are possible if $A \leq \mu \leq B$; the exact values of μ for which noise-induced transitions occur depend on the intensity of the additive noise.

Additive noise-induced switching has been detected in a variety of neural preparations. Examples include the switching times in the perception of ambiguous figures [Borsellino et al., 1972, Kruse and Stadler, 1995] and the detection of weak sensory signals via the mechanism of stochastic resonance [Gammaitoni et al., 1998, Moss et al., 1994]. The presence of three scaling regions in the two-point correlation function for the fluctuations in the center of pressure for postural sway has been suggested to reflect noise-induced switching between two periodic attractors [Eurich and Milton, 1996]. Empirical evidence for spontaneous switching between attractors in the brains of epileptic patients has been obtained from intracranial EEG recordings [Manuca et al., 1998]. In these studies, time series analysis suggested that the time variation of the EEG signals could be characterized by changes in a single variable. The observations were most consistent with a model of bistability in which mesoscopic collections of neurons flip between two collective states.

The second way that noise can enter into Equation 1 is in a state-dependent, or parametric, manner (Section 6), i.e.

$$\dot{V}(t) = f(V(t), V(t-\tau), \mu(\xi_P(t))), \tag{3}$$

where $\xi_P(t)$ is, for example, Gaussian distributed white noise. With respect to Figure 7.2, parametric noise produces horizontal fluctuations. Considerations of the effects of parametric noise are particularly relevant for neurobiologists since neural noise is typically state-dependent. For example, membrane noise reflects fluctuations in conductance [Verveen and DeFelice, 1974]. Current is proportional to the product of conductance and driving potential: hence the effect of the noise is state-dependent. State-dependent noise lies at the basis of the spontaneous fluctuations in pupil size [Stark et al., 1958] and in the clamped pupil light reflex [Milton et al., 1989]. It also plays an important role in motor [Harris and Wolpert, 1998] and balance [Cabrera and Milton, 2002] control.

Understanding the effects of parametric noise is more complicated than for additive noise because the full bifurcation diagram changes. Many of the same phenomena that have been observed in the presence of additive noise have their counterparts in the presence of parametric noise, for example, postponement of Hopf bifurcations [Longtin, 1991, Longtin et al., 1990]. Often the effects of parametric noise are more dramatic than those obtained with additive: additive noise can destroy a pitchfork bifurcation, parametric noise can destroy a Hopf bifurcation [Arnold et al., 1999]. However there are important differences between the effects of additive and parametric noise. Some of these differences are qualitative in nature: for noise-induced switching between attractors due to additive noise it must be true that $A \leq \mu \leq B$; for parametric noise the switching can occur even if, for example, $\mu < A$. More importantly parametric noise can produce phenomena that cannot be produced by additive noise. One such noise-induced phenomena is on-off intermittency discussed in Section 6.

The presence of multistability suggests that $A \leq \mu \leq B$. In Section 6 we show that measurements of the dynamics of neural networks lead to the surprising conclusion that for many networks, μ is tuned very near the critical point B (the $*$ in Figure 7.2). In Section 7 the possibility that self-organized criticality is a fundamental property of the nervous system is introduced [Beggs and Plenz, 2002, Chialvo and Bak, 1999, Kelso, 1999]. Thus changes in neural populations at seizure onset may share features similar to those that arise in phase transitions of physical systems [Horsthemenke and Lefever, 1984]. This observation has important implications for the design of an effective brain defibrillator.

3. Changes in neural synchrony

Epileptic seizures involve the synchronization of large populations of neurons [Jasper, 1969]; de-synchronization should abort the seizure. In this context a stimulating electrode produces an electric field that couples to neurons to different degrees related, in part, to each neuron's physical distance from it. This coupling serves to differentially stimulate the neurons and hence desynchronize the population [Durand, 1993]. Sensory stimuli could operate in a similar way by changing the activity in subcortical nuclei that in turn project diffusely to cortical neurons, e.g. the synchronizing effects of thalamo-cortical circuits [Contreras et al., 1996].

Perhaps the best understood mechanism that gives rise to synchronization is phase locking of oscillators. If a population of regularly firing neurons receives input from a periodic source, that population may become entrained, or synchronized, to that input if the frequencies are close, or are integer multiples of one another (n:m phase locking). The periodic input may arise from brain rhythms, such as hippocampal theta oscillations, that modulate the firing probability of neurons, or from periodic synaptic input to the population.

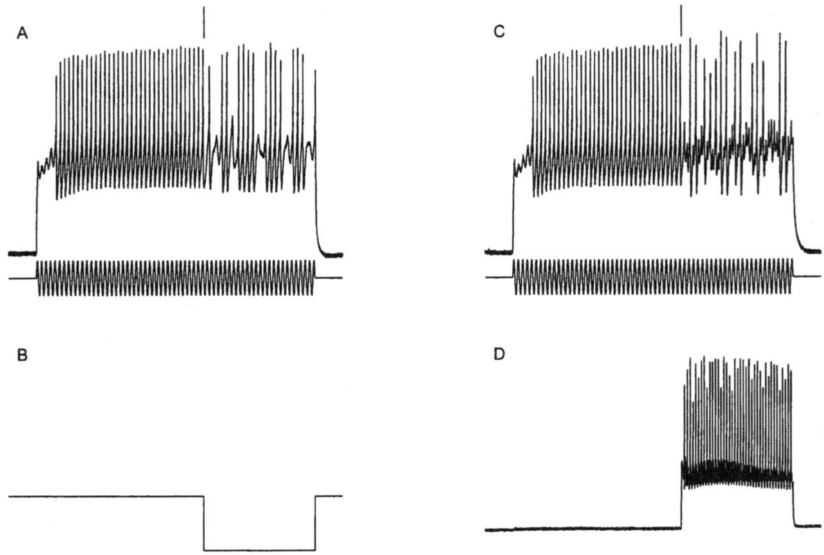

Figure 7.3. Neuronal rate changes produce rapid and large changes in entrainment. A) An *Aplysia* motoneuron (upper panels) is stimulated with a sinusoidal current which induces 1:1 phase locking. B) When a small hyperpolarizing current is applied, motoneuron firing is slowed and no longer is entrained in the 1:1 pattern with the sinusoid. Note that although a higher order locking pattern appears in the motoneuron response during the inhibitory phase, such patterns are less effective at generating population synchrony than a 1:1 pattern [Knight, 1972, Hunter and Milton, 2003]. C) The same motoneuron as in A. D) In place of the hyperpolarizing current, an inhibitory interneuron is activated which is presynaptic to the motoneuron. This too, slows motoneuron firing resulting in a nearly instantaneous loss of synchrony.

Since the ability of the population to synchronize to the input is a function of the firing rate of the population, small changes in firing rate can induce large changes in neuronal synchronization [Hunter et al., 1998, Hunter and Milton, 2002, Hunter and Milton, 2003]. Figure 7.3 compares the effect on synchronization to a periodic input between two different sources of rate modulation: 1) tonic current injection such as one might find with a seizure focus stimulating electrode and 2) synaptic stimulation, which might arise when the target site is separated from the stimulus site via synapses as in the case of the vagal nerve stimulator.

Figure 7.3A shows a motoneuron from the *Aplysia* buccal ganglion which is stimulated by a sine wave with 1:1 phase locking. As such, the neuron is highly entrained to the sine wave. At 3.5s, a hyperpolarizing current is applied to the motoneuron, slowing motoneuron firing and reducing entrainment. Thus tonic current injection is a means to sensitively control population entrainment to a coherent periodic input. The change in synchrony is nearly instantaneous. While this is certainly a desirable feature in a seizure control methodology, it does not accord with observations such as those in Figure 7.1A, where the response to a perturbation occured more than 1s after the stimulation.

In Figure 7.3B, we use a synaptic input to cause changes in motoneuron firing rate. As in Figure 7.3A, the change in motoneuron entrainment to the periodic input is nearly instantaneous. In some cases, the motoneuron shows evidence of entrainment to the interneuron as well [Hunter and Milton, 2003], pointing to a potential problem of trying to induce desynchronization via a coherent input to a neural population.

These observations suggest that changes in neural synchrony to electrical and synaptic inputs typically occur rapidly. Thus this mechanism does not readily explain the magnitude of the latency observed between stimulus onset and seizure cessation.

4. Multistability in delayed recurrent loops

Recurrent inhibitory loops play an important role in epileptic seizures arising from the amygdala-hippocampal complex [Schwartzkroin and McIntyre, 1997] and those generated from thalamocortical interactions [Coulter, 1997] (Figure 7.4A): an excitatory neuron, E, gives off collateral branches that excite an inhibitory interneuron, I, which in turn, inhibits the firing of E [Mackey and an der Heiden, 1984, Mackey and Milton, 1987, Milton, 1996, Milton et al., 1990]. Time delays arise in recurrent loops because of the time taken for the inhibitory signal to traverse the recurrent loop. Consequently, mathematical models take the form of delay differential equations (DDE). It has been shown that multistability readily arises in mathematical models of delayed neural recurrent loops [Foss et al., 1996, Foss et al., 1997b] and experimentally in electrical circuits [Foss et al., 1997b] and in recurrently clamped neurons [Foss and Milton, 2000].

The complexities that arise in the use of stimuli to cause switches between attractors in delayed recurrent loops can be appreciated by considering a simple integrate-and-fire approximation to the recurrent inhibitory loop (Figure 7.4B). The membrane potential of a neuron increases linearly until it reaches the firing threshold at which point the neuron fires, and the membrane potential is reset to its resting value. The firing of the neuron excites the inhibitory neuron, I, which in turn, at a time τ later, delivers an inhibitory pulse to the excitatory neuron, E. The advantage of this simple model is that in dimensionless form, the dynamics depend only on two parameters: the magnitude of the inhibitory pulse, Δ, and τ. Consequently considerable insight can be obtained into the behavior of the loop in response to perturbations, including noise.

The solutions of this integrate-and-fire delayed recurrent loop can be constructed from segments of length τ; each segment satisfies an equation of the

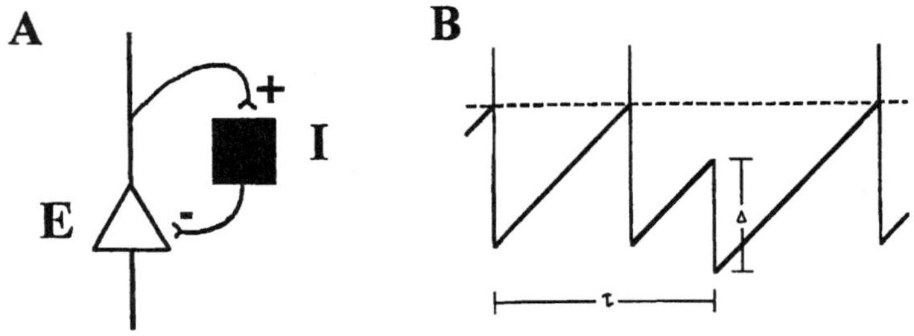

Figure 7.4. A) Schematic representation of a recurrent inhibitory neural loop. B) The time course for the membrane potential (vertical axis) for an integrate-and-fire approximation for the recurrent loop shown in A). The horizontal dashed line is the firing threshold and the time delay, τ, is the time taken for neural activity to traverse the recurrent loop [Foss et al., 1997b, Milton and Foss, 1997].

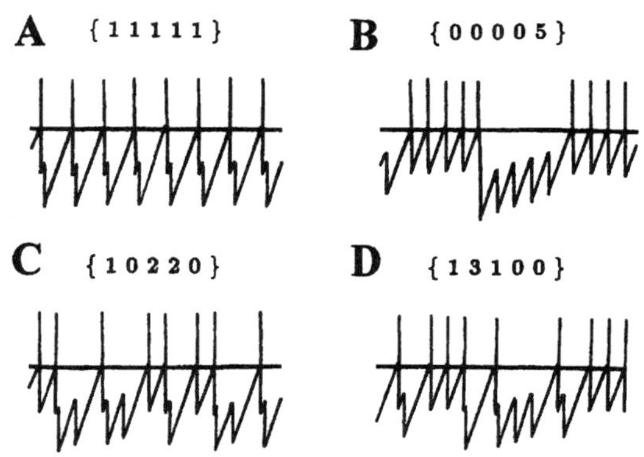

Figure 7.5. The four attractors that co-exist in the integrate-and-fire recurrent inhibitory loop when $\tau = 4.1$ and $\Delta = 0.8$.

form

$$\tau = x + m + n\Delta \tag{4}$$

where n, m are positive integers and $0 < x < 1$ [Foss et al., 1997b]. For τ, Δ fixed, the total number of (m, n) pairs that satisfy Equation 4 is the smallest integer greater than τ/Δ. Since the number of such (m, n) pairs is finite it follows that all solutions are periodic. When $\tau < 1$, only a single attractor exists with period $1 + \Delta$. Multistability occurs when $\tau > 1$. This is because the inhibitory pulses are not necessarily the result of the preceding excitatory spike. Figure 7.5 shows that four qualitatively different periodic attractors exist when $\tau = 4.1$ and $\Delta = 0.8$: $\{11111\}$, $\{00005\}$, $\{10220\}$, $\{13100\}$, where the notation gives the repeating unit of the periodic solution described by the number of inhibitory pulses per interspike interval.

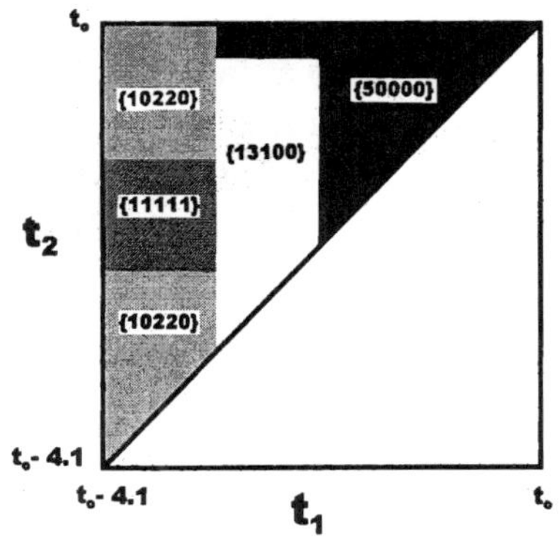

Figure 7.6. Structure of the functional space, ψ_3, for the integrate-and-fire recurrent loop with $\tau = 4.1, \Delta = 0.8$. This functional space is symmetric about the 45deg line; we show only the upper half.

It is not straightforward to understand how a switch between two attractors occurs in a DDE that exhibits multistability. To appreciate the complexities it is important to note that a DDE can be thought of as a procedure that maps a function, ψ, of length τ onto a function of length τ. For the integrate-and-fire recurrent loop, the functions, ψ, are specified by the times of spike occurrence. For $\tau = 4.1$, $\Delta = 0.8$, a sequence of three spike times, $\psi_3 = \{t_1, t_2, t_3\}$ is sufficient to initialize the four coexistent attractors [Milton and Foss, 1997]. Each ψ leads to a unique solution. As shown in Figure 7.6, the space of all possible ψ_3 is partitioned between the four attractors. The integrate-and-fire recurrent loop evolves in a functional space constructed from possible $\psi_i, i = 1,, \cdots, n$.

In general for multistable DDEs, the structure of these basins of attraction in ψ-space is exceedingly complex (for examples, see [Bayer and an der Heiden, 1998, Foss et al., 1996, Losson et al., 1993, Milton and Foss, 1997]. In order to understand the effect of introduced perturbations it is convenient to think of ψ as being composed of two components: one component represents the spike pattern generated as a solution of the delayed recurrent loop; the other component represents the perturbations introduced by the experimentalist (or noise, see next section) in an attempt to induce a switch between attractors. The problem faced by the experimentalist is to add precisely timed perturbations in order to change ψ into a new ψ that maps to a new basin of attraction. In general it will not be possible to accomplish this task by adding a single pulse (as occurs in dynamical systems that lack retarded variables [Guttman et al., 1980, Winfree, 1980]).

Figure 7.7 shows an example in which the introduction of a single, carefully timed inhibitory pulse causes a switch between two qualitatively different spiking patterns. The pattern of spiking between the regions designated $\{11111\}$ and $\{13100\}$ does not correspond to a stable solution of the model and hence represents a transient. This transient lasts $\sim 2\tau$. Switching times longer than τ are also observed between attractors in a recurrently clamped *Aplysia* motoneuron (Figure 7.8); here the switching time is between τ and 2τ.

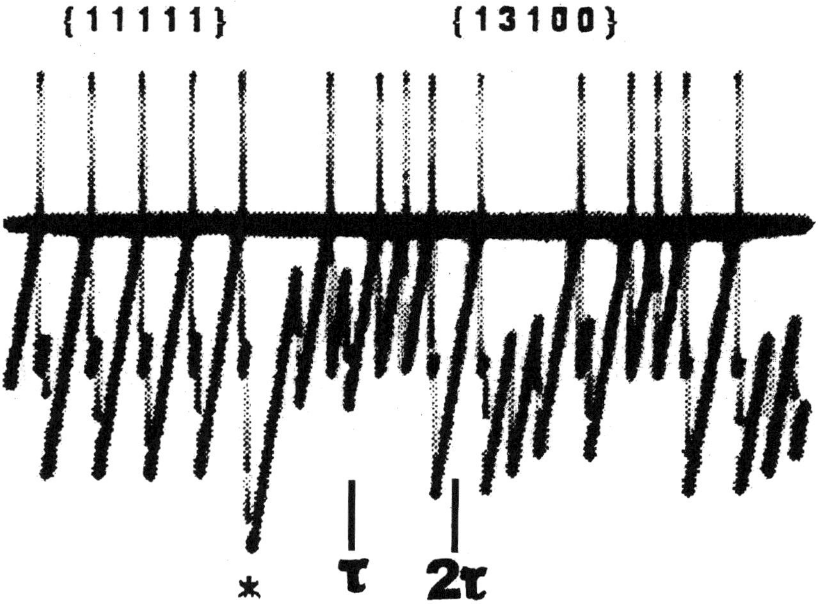

Figure 7.7. When $\tau = 4.1, \Delta = 0.8$, a carefully timed inhibitory pulse ($*$) induces a switch from the $\{11111\}$-attractor to the $\{13100\}$-attractor. This switching is shown for an electrical circuit that mimics the integrate-and-fire recurrent loop shown in Figure 7.4 [Foss et al., 1997b]. The time after stimulus presentation is shown in terms of τ.

The fact that the switching time is longer than τ is easy to appreciate: a function of length τ is required to define an attractor. One reason that switching times longer than τ arise is because of transients that arise as trajectory settles down onto the new attractor. In the next section we show that a switch between basins of attraction typically requires a number of steps.

5. Noise-induced switching between attractors

The advantage of studying noise-induced transitions between two attractors is that it provides insights into the basic underlying mechanism of how such changes occur. Insights into how switches occur in dynamical systems that possess retarded variables can be readily obtained from the integrate-and-fire recurrent loop. Since in this model only the timing of spikes is important, it is

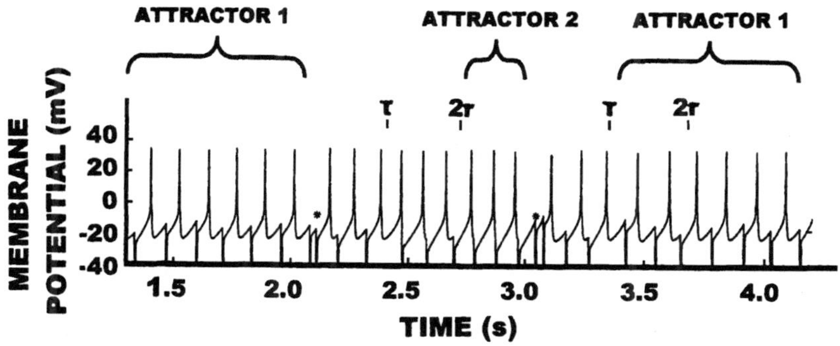

Figure 7.8. When $\tau = 315$ms a carefully timed inhibitory pulse ($*$) induces a switch from one attractor to another and then back again in a neural loop constructed from a recurrently clamped *Aplysia* motoneuron [Foss and Milton, 2000]. The time after stimulus presentation is shown in terms of τ. Clearly it is difficult to determine the exact point in time at which a switch occurs between the two attractors and whether or not transients are present.

equivalent to inject noise into τ or Δ. In the discussion that follows we consider the case that noise has been added to Δ.

Two steps are involved in producing a switch between two attractors: 1) leaving the basin of attraction of the first attractor; and 2) entering the basin of attraction of the second. The exiting step can be assessed through measurements of the dwell time. The dwell time is defined as the time interval between when a trajectory first enters a given basin of attraction to the time that it leaves. In the case of the integrate-and-fire recurrent loop a trajectory is said to leave a basin of attraction when the number of inhibitory pulses in an interspike interval differs from that expected. It has been shown that the fraction of trajectories remaining in a given basin of attraction decreases exponentially as a function of time in the presence of noise [Foss et al., 1997b]. An exponential distribution of dwell times is the characteristic distribution observed for the times to cross a threshold in a stochastic [Kramers, 1940] or a chaotic [Bauer and Bertsch, 1990, Legrand and Sornette, 1990, Mackey and Milton, 1990] dynamical system. An exponential distribution of times between seizures is observed for patients with medically intractable epilepsy who take anti-convulsant medications [Milton et al., 1987].

A more complex question is determining what happens to the trajectory once it has exited the basin of attraction: Does it eventually return to the original attractor? Does it switch to new attractor? Typically the length of transients is long compared to τ, and the time between successive noisy perturbations is short compared to τ. Thus noise-prolonged transients can exist for considerable periods of time and represent an example of metastability [Grotta-Ragazzo et al., 1999].

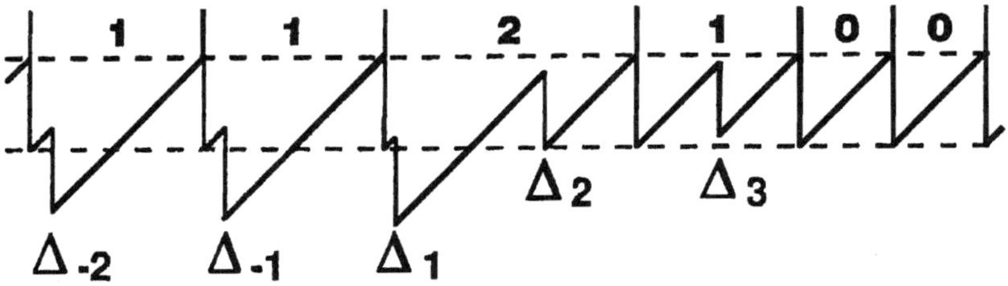

Figure 7.9. Schematic representation of the most common path from the $\{11111\}$-attractor to the $\{13100\}$-attractor [Foss et al., 1997b].

Figure 7.9 shows the most common path by which a switch occurs between the $\{11111\}$-attractor and the $\{13100\}$-attractor in the delayed integrate-and-fire recurrent loop [Foss et al., 1997b]. The positive subscripts refer to those Δ's that occur after the trajectory leaves the attractor; the negative subscripts refer to those that occur prior to this event. Thus the transients involved in the switch between two attractors in this model reflect, in part, the fact that three carefully timed Δ's are required to effect the switch.

In general there are an arbitrarily large number of paths from one attractor to another. In the delayed integrate-and-fire recurrent loop, most of these paths have a negligible probability of occurring. This is because of the constraints placed on more than five Δ's and the requirement for many paths for the value of Δ to be more than three standard deviations from the mean.

However, the situation becomes much more complicated in models with continuous feedback [Grotta-Ragazzo et al., 1999]. For example, the condition for multistability in the Mackey-Glass equation [Mackey and Glass, 1977], i.e.

$$\dot{x}(t) + \alpha x(t) = \frac{\beta \Upsilon^n x(t-\tau)}{\Upsilon^n + x(t-\tau)^n}, \tag{5}$$

is that τ be greater than a^{-1}, where a, n, β, Υ are constants. A particularly convenient case to study noise-induced switching between two attractors is when $n = 8$ [Foss et al., 1997a]: one attractor exists for $x > 0$, another for $x < 0$ [Losson et al., 1993]. In this case with additive noise numerical simulations demonstrate that the switching times between the two co-existent basins of attractors are non-exponential and possess long tails with switching times exceeding 1000τ [Foss et al., 1997a]! Thus even if time delays are of the order of 10 ms, switching times of the order of seconds are possible. Non-exponential survival statistics characterize the response of after-discharges to brief electrical pulses [Lesser et al., 1999].

6. On-off intermittency: Parametric, or state-dependent, noise

The effects of parametric noise on the dynamics of neural recurrent loops have not yet been studied. However, recently important insights into the effects of parametric noise on neural control have been obtained from studies of stick balancing at the fingertip [Cabrera and Milton, 2002, Cabrera and Milton, 2003, Mehta and Schaal, 2002]. This visuomotor task cannot be performed using memorized movement patterns, but requires continuous closed-loop control. This paradigm has a number of advantages that make it well suited to explorations of neural dynamics: 1) the movements of the balanced stick can be measured with high precision using non-invasive 3-D motion analysis techniques; and 2) mathematical models for the control of the inverted pendulum with time-delayed feedback have been developed previously [Stépàn, 1989]. Thus it is possible to directly compare prediction with experimental observation.

The crucial role played by time-delayed feedback in stick balancing is demonstrated by the observation that longer sticks are much easier to balance than shorter ones. This is because once the stick is sufficiently long, its rate of movement becomes slow relative to the time required to make corrective movements. The noisy fluctuations enter by the movements of the fingertip (hand) that act through changes in the pivot point of the inverted pendulum, i.e. as a parametric perturbation [Bogdanoff, 1962, Bogdanoff and Citron, 1965]. Even for slow moving sticks the effects of neural feedforward predictive control is small: trained subjects are only able to maintain control throughout visual blank-out periods that last $< 2 - 3\tau$, where τ corresponds to the neural latency for stick balancing [Mehta and Schaal, 2002].

The controlled variable is the vertical displacement angle of the balanced stick, θ. Figure 7.10A shows the fluctuations in $\Delta z/l$ (equal to $\cos\theta$) as a function of time, where l is the length of the stick and Δz is the difference in the vertical coordinate of the upper and lower ends of the stick. These fluctuations exhibit intermittency i.e. intermittently there are large deviations. Dynamical systems that exhibit intermittency are often characterized by the presence of power laws. Figure 7.10B shows that the laminar phases exhibit a $-3/2$ power law. The term *laminar phase* refers to the time interval between successive large fluctuations. The laminar phases were measured by first choosing an arbitrary threshold (solid horizontal line in Figure 7.10B), and then, measuring the time intervals between successive threshold crossings in the upward direction (i.e. successive corrective movements). The power spectrum of the fluctuations in θ (not shown) contains two regions of $1/f^{\alpha}$ behavior: one with $\alpha \sim 0.5$, another at higher frequencies with $\alpha \sim 2.5$. A $-3/2$ power law (together with a $-1/2$

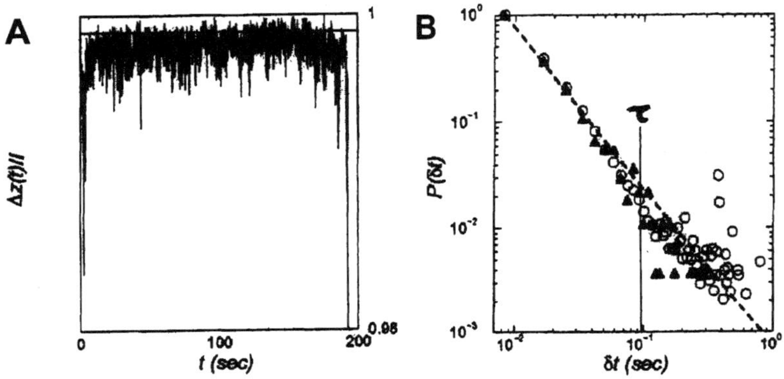

Figure 7.10. A) Temporal series for $\Delta z/l$ (equal to $\cos\theta$) for a 62 cm stick balanced at the fingertip. The horizontal line depicts the threshold position at 1.005 times the mean value. B) Log-log plot of the probability that the time between successive corrective movements for stick balancing at the fingertip is longer than δt, $P(\delta t)$. The latency for stick balancing is $\sim$ 100 ms for this subject. Data from [Cabrera and Milton, 2002].

power law in power spectra [Venkataramani et al., 1996]) strongly suggests that the fluctuations in θ exhibit on-off intermittency.

In deterministic dynamical systems, Type III intermittency produces laminar phases with a $-3/2$ power law [Pomeau and Manneville, 1980]. In stochastic dynamical systems on-off intermittency arises from the stochastic or chaotic forcing of a control parameter across a stability boundary [Ding and Yang, 1995,Heagy et al., 1994,Platt et al., 1993]. This power law cannot be reproduced using additive noise; it only is seen in the presence of parametric noise.

The important corollary of the observation of on-off intermittency in a noisy dynamical system is that its presence implies that the control mechanism must be tuned in parameter space to be close, or perhaps on, a stability boundary. Power laws are expected to arise when control mechanisms are tuned at stability boundaries [Bak et al., 1988, Chialvo and Bak, 1999, Kadanoff, 1993, Stanley et al., 1998]. Indeed it has been shown that the changes in speed made by the hand during stick balancing are Lévy distributed and follow a power law [Cabrera and Milton, 2003]. These studies suggest that the neural control of stick balancing exhibits self-organized criticality.

It has been suggested that self-organized criticality is a feature of neural networks that learn by making mistakes [Chialvo and Bak, 1999]. However, the advantages of self-organized criticality for neural control are not yet clear. One possible advantage is that it enables balance control to be maintained on time scales shorter than the delay [Cabrera and Milton, 2002]. Indeed, on examining Figure 7.10B we see that $> 98\%$ of the time intervals between successive corrective movements are shorter than the neural latency for stick balancing. In particular, the fluctuations in θ resemble a random walk for which the mean value of θ is approximately zero, i.e. the upright position is "statistically stabilized" [Cabrera and Milton, 2002].

7. Self-organized criticality

Two recent observations suggest that self-organized criticality is not just restricted to the control of stick balancing, but may be a fundamental organizing property of neural populations. First, the bursting activity of neurons grown in culture exhibits scaling phenomena [Segev et al., 2002]. Secondly, and more relevant for our discussion, cortical networks in slices of rat cortex have been shown to spontaneously evolve into an activity regime that lies on a boundary between the extremes of complete randomness and redundant order [Beggs and Plenz, 2002]. In these experiments, spontaneously occurring fluctuations in the negative peaks of local field potentials were measured in isolated slices of rat cortex using microelectrode arrays. Using the same approach for data analysis employed for stick balancing, an arbitrary threshold was chosen and the time between successive threshold crossings measured. It was found that the laminar phases exhibited a $-3/2$ power law. These authors speculate that cortical neural networks exist on the edge between epileptic seizures on the one hand, and quiescence on the other. At the very least these observations provide a nice explanation of why intrinsic neural mechanisms to stop seizures exist even in the non-epileptic brain [Chkhenkeli, 2002, Chkhenkeli and Milton, 2002].

8. Discussion

Is epilepsy an example of a dynamic disease [Milton and Black, 1995, Milton and Jung, 2002]? A dynamic disease is defined as a disease that occurs in an intact physiological control system that operates in a range of control parameters [Glass and Mackey, 1979, Mackey and Glass, 1977, Mackey and Milton, 1987]. In other words the pathology of the disease is dynamical rather than structural. It is certainly true that the occurrence of an epileptic seizure represents a qualitative change in the dynamics of neuronal activity. Qualitative changes in dynamics are one of the hallmarks of a dynamic disease. If epilepsy is a dynamic disease, then its treatment should be dynamic, i.e. therapeutic strategies must be based on manipulation of the underlying dynamics. The hope is that these dynamic strategies can be identified and eventually implemented as the treatment arm of a brain defibrillator.

In dynamical systems with multiple basins of attraction, changes in dynamics can arise as a consequence of: 1) perturbations that take you from one basin to the other; 2) changes in the basins of attraction and their boundaries due to changes in control parameters; and 3) changes in control parameters that destabilize the co-existent attractors. Since the control parameters whose change heralds seizure onset have not yet been identified we have focused on the first two mechanisms for causing switches between basins of attraction. From the point of view of the fixed-point attractor in Figure 7.2, the presence of multistability implies that the control parameters must be tuned close to the edge of its instability. *The identification of critical phenomena (see below) in neural dynamics strongly implies that the control parameters are not just tuned close to the edge of instability, but in fact are tuned at the edge of instability!* In other words, neural dynamical networks exhibit self-organized criticality [Kelso, 1999].

Briefly timed electrical stimuli are also used to control seizures based on "control of chaos" techniques [Spano et al., 2002, Schiff et al., 1994]. The application of this technique is based on the identification of unstable periodic orbits in a time series by the presence of so-called "walking in, walking out" phenomena. It is important to realize the underlying dynamic structure for this phenomenon, namely a saddle node bifurcation, is precisely the same as that of the separatrix that separates two basins of attraction in a multistable dynamical system. It has been suggested that it may be very difficult to distinguish between a chaotic dynamical system and a noisy multistable dynamical system [Glanz, 1997]. However, a chaotic attractor exists within a basin of attraction, whereas the presence of multistability requires that the dynamical system be tuned close to a stability boundary.

In the presence of noise, dynamical systems tuned close to instability are anticipated to exhibit a number of critical phenomena. There is growing evidence that critical phenomena, in addition to multistability, can indeed be detected in neural dynamics: power laws [Collins and De Luca, 1994, Fitts and Posner, 1973], on-off intermittency [Beggs and Plenz, 2002, Cabrera and Milton, 2002], noise-induced postponement of bifurcations [Baer et al., 1989, Longtin, 1991, Longtin et al., 1990, Rinzel and Baer, 1988]; critical slowing down [Longtin and Hinzer, 1996, Matsumoto and Kunisawa, 1978], noise amplification [Longtin et al., 1990], phase transitions [Haken et al., 1985, Kelso, 1984, Kelso et al., 1992, Meyer-Lindenberg et al., 2002, Schöner et al., 1986], scale-invariant Lévy phenomena [Brockman and Giesel, 2000, Cabrera and Milton, 2003, Segev et al., 2002, Viswanathan et al., 1996]. Moreover it has been demonstrated that self-organized criticality can arise in plausible neural networks that learn by making mistakes [Chialvo and Bak, 1999].

Traditionally the study of patients with epilepsy has provided the greatest insights into the nature of the central nervous system of humans [Jackson, 1931, Penfield and Jasper, 1954, Morrell, 1985]. The identification of critical phenomena in relation to phase transitions (e.g. the transition from liquid water to ice) motivated a shift in attention of physicists away from considerations of single atoms and molecules to an emphasis on the behavior of networks of molecules [Barabási, 2002, Kadanoff, 1993]. Is it possible that a similar revolution is about to occur in our understanding of the dynamics of neural networks? Do partial complex seizures arise from an epileptic focus or do they represent the emergent dynamics of a large distributed epileptic network [Chkhenkeli and Milton, 2002] that arise via phase transitions? Thus the search for a therapeutic strategy to treat epileptic seizure appears to lie directly on the route that may uncover the basic organizational structure(s) of large neural networks. All that will be required is a dash of clinical epilepsy mixed with a sprinkle of basic neuroscience and a dab of physics!

Acknowledgments

We thank S. A. Campbell, C. W. Eurich, N. G. Hatsopoulos, L. P. Kadanoff, M. C. Mackey, R. McCrea and T. Ohira for helpful discussions. This work was supported by grant from the Brain Research Foundation (BRF) and NIMH. JLC acknowledges support from MECD (Spain) and BRF.

References

[Ajmone Marsan, 1972] Ajmone Marsan, C. (1972). Focal electrical stimulation. In Purpura, D. P., Penry, J. K., Tower, D. B., Woodbury, D. M., and Walter, R. D., editors, *Experimental models of epilepsy: A manual for the laboratory worker*, pages 147–172, New York. Raven Press.

[an der Heiden, 1979] an der Heiden, U. (1979). Delays in physiological systems. *J. Math. Biol.*, 8:345–364.

[an der Heiden and Mackey, 1982] an der Heiden, U. and Mackey, M. C. (1982). The dynamics of production and destruction: Analytic insight into complex behavior. *J. Math. Biol.*, 16:75–101.

[Arielle et al., 1996] Arielle, A., Sterkin, A.and Grinvald, A., and Aerstein, A. (1996). Dynamics of ongoing activity: Explanation of the large variability in evoked potential responses. *Science*, 273:1868–1871.

[Arnold, 1998] Arnold, L. (1998). *Random dynamical systems*. Springer-Verlag, New York.

[Arnold et al., 1999] Arnold, L., Bleckert, G., and Schenk-Hoppé, K. (1999). The stochastic brusselator: Parametric noise destroys Hopf bifurcation. In Crauel, H. and Gundlach, M., editors, *Stochastic Dynamics*, pages 71–92. Springer-Verlag, New York.

[Baer et al., 1989] Baer, S. M., Erneux, T., and Rinzel, J. (1989). The slow passage through a Hopf bifurcation: Delay, memory effects, and resonance. *SIAM J. Appl. Math.*, 49:55–71.

[Bak et al., 1988] Bak, P., Tang, C., and Wiesenfeld, K. (1988). Self-organized criticality. *Phys. Rev. A*, 38:364–374.

[Barabási, 2002] Barabási, A.-L. (2002). *Linked: The new science of networks*. Perseus Publishing, Cambridge, MA.

[Bauer and Bertsch, 1990] Bauer, W. and Bertsch, G. F. (1990). Decay of ordered and chaotic systems. *Phys. Rev. Lett.*, 65:2213–2216.

[Bayer and an der Heiden, 1998] Bayer, W. and an der Heiden, U. (1998). Oscillation types and bifurcations of a nonlinear second-order differential-difference equation. *J. Dyn. Diff. Eqns.*, 10:303–326.

[Beggs and Plenz, 2002] Beggs, J. M. and Plenz, D. (2002). Self-organized criticality of spontaneous activity in isolated cortical networks. *Society of Neuroscience (Abstracts)*, page 28.

[Bogdanoff, 1962] Bogdanoff, J. L. (1962). Influence on the behavior of a linear dynamical system of some imposed rapid movements of small amplitude. *J. Acoust. Soc. Amer.*, 34:1055–1062.

[Bogdanoff and Citron, 1965] Bogdanoff, J. L. and Citron, S. J. (1965). Experiments with an inverted pendulum subject to random parametric perturbations. *J. Acoust. Soc. Amer.*, 38:447–452.

[Borsellino et al., 1972] Borsellino, A., De Marco, A., Allazetta, A., Rinesi, S., and Bartolini, B. (1972). Reversal time distributions of visual ambiguous stimuli. *Kybernetik*, 10:139–144.

[Brockman and Giesel, 2000] Brockman, D. and Giesel, T. (2000). The ecology of gaze shifts. *Neurocomputing*, 32-33:643–650.

[Cabrera and Milton, 2002] Cabrera, J. L. and Milton, J. G. (2002). On-off intermittency in a human balancing task. *Phys. Rev. Lett.*, 89:158702–1–4.

[Cabrera and Milton, 2003] Cabrera, J. L. and Milton, J. G. (2003). Delays, scaling and the acquisition of motor skill. In Bezrukov, S., editor, *Unsolved Problems of Noise and Fluctuations: UpoN 2002: Third International Conference on Unsolved Problems of Noise and Fluctuations in Physics, Biology and High Technology* (AIP Proceedings Vol. 665), pages 250–256, Melville, NY. American Institute of Physics.

[Campbell et al., 1995] Campbell, S. A., Bélair, J., Ohira, T., and Milton, J. G. (1995). Limit cycles, tori and complex dynamics in a second-order differential equation with delayed negative feedback. *J. Diff. Eqns.*, 7:213–236.

[Chialvo and Bak, 1999] Chialvo, D. R. and Bak, P. (1999). Learning from mistakes. *Neuroscience*, 90:1137–1148.

[Chkhenkeli, 2002] Chkhenkeli, S. A. (2002). Direct deep brain stimulation: First steps towards the feedback control of seizures. In Milton, J. and Jung, P., editors, *Epilepsy as a dynamic disease*, pages 249–262, New York. Springer-Verlag.

[Chkhenkeli and Milton, 2002] Chkhenkeli, S. A. and Milton, J. (2002). Dynamic epileptic systems versus static epileptic foci? In Milton, J. and Jung,

P., editors, *Epilepsy as a dynamic disease*, pages 25–36, New York. Springer-Verlag.

[Collins and De Luca, 1994] Collins, J. J. and De Luca, C. J. (1994). Random walking during quiet standing. *Phys. Rev. Lett.*, 73:764–767.

[Contreras et al., 1996] Contreras, D., Destexhe, A., Sejnowski, T. J., and Steraide, M. (1996). Control of spatiotemporal coherence of a thalamic oscillaton by corticothalamic feedback. *Science*, 274:771–774.

[Coulter, 1997] Coulter, D. A. (1997). Thalamocortical anatomy and physiology. In Engel, Jr., J. and Pedley, T. A., editors, *Epilepsy: A comprehensive textbook*, pages 341–351, Philadelphia. Lippincott-Raven.

[Ding and Yang, 1995] Ding, M. and Yang, W. (1995). Distribution of the first return time in fractional Brownian motion and its application to the study of on-off intermittency. *Phys. Rev. E*, 52:207–213.

[Durand, 1993] Durand, D. M. (1993). Ictal patterns in experimental models of epilepsy. *Clin. Neurophysiol.*, 10:281–297.

[Ermentrout and Kopell, 1994] Ermentrout, B. B. and Kopell, N. (1994). Learning of phase lags in coupled neural oscillators. *Neural Comp.*, 6:225–241.

[Eurich and Milton, 1996] Eurich, C. W. and Milton, J. G. (1996). Noise-induced transitions in human postural sway. *Phys. Rev. E*, 54:6681–6684.

[Fitts and Posner, 1973] Fitts, P. M. and Posner, M. I. (1973). *Human performance*. Prentice-Hall, London.

[Foss et al., 1997a] Foss, J., Eurich, C. W., Milton, J., and Ohira, T. (1997a). Noise, multistability and long-tailed interspike interval (ISI) histograms. *Bull. Amer. Phys. Soc.*, 42:781.

[Foss et al., 1996] Foss, J., Longtin, A., Mensour, B., and Milton, J. (1996). Multistability and delayed recurrent loops. *Phys. Rev. Lett.*, 76:708–711.

[Foss and Milton, 2000] Foss, J. and Milton, J. (2000). Multistability in recurrent neural loops arising from delay. *J Neurophysiology*, 84:975–985.

[Foss and Milton, 2002] Foss, J. and Milton, J. (2002). Aborting seizures with a single stimulus: The case for multistability. In Milton, J. and Jung, P., editors, *Epilepsy as a dynamic disease*, pages 283–295, New York. Springer-Verlag.

[Foss et al., 1997b] Foss, J., Moss, F., and Milton, J. (1997b). Noise, multistability, and delayed recurrent loops. *Phys. Rev. E*, 55:4536–4543.

[Gammaitoni et al., 1998] Gammaitoni, L., Hänggi, P., Jung, P., and Marchesoni, F. (1998). Stochastic resonance. *Rev. Mod. Phys.*, 70:223–288.

[Glanz, 1997] Glanz, J. (1997). Mastering the nonlinear brain. *Science*, 277:1758–1760.

[Glass and Mackey, 1979] Glass, L. and Mackey, M. C. (1979). Pathological conditions resulting from instabilities in physiological control systems. *Ann. N. Y. Acad. Sci.*, 620:22–44.

[Gotman, 1983] Gotman, J. (1983). Measurement of small time differences between EEG channels method and application to epileptic seizure propagation. *Electroenceph. Clin. Neurophysiol.*, 56:501–514.

[Grotta-Ragazzo et al., 1999] Grotta-Ragazzo, C., Pakdaman, K., and Malta, C. P. (1999). Metastability for delayed differential equations. *Phys. Rev. E*, 60:6230–6233.

[Guillouzic et al., 1999] Guillouzic, S., L'Heureux, I., and Longtin, A. (1999). Small delay approximation of stochastic delay differential equations. *Phys. Rev. E*, 59:3970–3982.

[Guttman et al., 1980] Guttman, R., S., L., and Rinzel, J. (1980). Control of repetitive firing in squid axon membrane as a model for a neuron oscillator. *J. Physiol.*, 305:377–395.

[Haken et al., 1985] Haken, H., Kelso, J. A. S., and Bunz, H. (1985). A theoretical model of phase transitions in human movement. *Biol. Cybern.*, 53:247–257.

[Harris and Wolpert, 1998] Harris, C. M. and Wolpert, D. M. (1998). State-dependent noise determines motor planning. *Nature (London)*, 394:780–784.

[Heagy et al., 1994] Heagy, J. F., Platt, N., and Hammel, S. M. (1994). Characterization of on-off intermittency. *Phys. Rev. Lett.*, 49:1140–1150.

[Hetling, 2002] Hetling, J. R. (2002). Prospects for building a therapeutic cortical stimulator. In Milton, J. and Jung, P., editors, *Epilepsy as a dynamic disease*, pages 323–339, New York. Springer-Verlag.

[Horsthemenke and Lefever, 1984] Horsthemenke, J. and Lefever, R. (1984). *Noise-induced transitions: Theory and applications in physics, chemistry and biology*. Springer-Verlag, New York.

[Hunter and Milton, 2001] Hunter, J. D. and Milton, J. G. (2001). Synaptic heterogeneity and stimulus induced modulation of depression in central synapses. *J. Neurosci.*, 21:1427–1438.

[Hunter and Milton, 2002] Hunter, J. D. and Milton, J. G. (2002). Using inhibitory interneurons to control neural synchrony. In Milton, J. G. and Jung, P., editors, *Epilepsy as a Dynamic Disease*, pages 115–130, New York, Springer-Verlag.

[Hunter and Milton, 2003] Hunter, J. D. and Milton, J. G. (2003). Amplitude and frequency dependence of spike timing: Implications for dynamic regulation. *J. Neurophysiology* 90:387–394.

[Hunter et al., 1998] Hunter, J. D., Milton, J. G., Thomas, P. J., and Cowan, J. D. (1998). Resonance effect for neural spike time reliability. *J. Neurophysiol*, 80(3):1427–38.

[Izhikevich, 2000] Izhikevich, E. M. (2000). Neural excitability, spiking, and bursting. *Int. J. Bifurc. Chaos*, 10:1171–1266.

[Jackson, 1931] Jackson, J. H. (1931). *Selected Writings*. Hodder & Soughton, London.

[Jasper, 1969] Jasper, H. H. (1969). Mechanisms of propagation: Extracellular studies. In Jasper, H. H., Ward, A. A., and Pope, A., editors, *Basic mechanisms of the epilepsies*, pages 421–440, Boston. Little Brown.

[Kadanoff, 1993] Kadanoff, L. P. (1993). *From order to chaos, essays: Critical, chaotic and otherwise*. World Scientific, Singapore.

[Kelso, 1984] Kelso, J. A. S. (1984). Phase transitions and critical behavior in human bimanual coordination. *Amer. J. Physiology: Regulation, integrative and comparative physiology*, 15:R1000–R1004.

[Kelso, 1999] Kelso, J. A. S. (1999). *Dynamical patterns: The self-organization of brain and behavior*. The MIT Press, Cambridge, MA.

[Kelso et al., 1992] Kelso, J. A. S., Bressler, S. L., Buchanan, S., DeGuzman, G. C., Ding, M., Fuchs, A., and Holroyd, T. (1992). A phase transition in human brain and behavior. *Phys. Lett. A*, 169:134–144.

[Knight, 1972] Knight, B. K. (1972). Dynamics of encoding in a population of neurons. *J Gen. Physiol.*, 59:734–766.

[Kopell, 1995] Kopell, N. (1995). Chains of coupled oscillators. In Arbib, M. A., editor, *Brain theory and neural networks*, pages 178–183, Cambridge, MA. MIT Press.

[Kramers, 1940] Kramers, H. A. (1940). Brownian motion in a field of force and the diffusion model of chemical reactions. *Physica*, 7:284–304.

[Kruse and Stadler, 1995] Kruse, P. and Stadler, M., editors (1995). *Ambiguity in Mind and Nature*. Springer-Verlag, New York.

[Lechner et al., 1996] Lechner, H. A., Baxter, D. A., Clark, J. W., and Byrne, J. H. (1996). Bistability and its regulation by serotonin in the endogenously bursting neuron R15 of *Aplysia*. *J. Neurophysiology*, 75:957–962.

[Legrand and Sornette, 1990] Legrand, O. and Sornette, D. (1990). Coarse-grained properties of the chaotic trajectories in the stadium. *Physica D*, 44:229–247.

[Lesser et al., 1999] Lesser, R. P., Kim, S. H., Beyderman, L., Miglioretti, D. L., Webber, W. R. S., Bare, M., Cysyk, B., Krauss, G., and Gordon, B. (1999). Brief bursts of pulse stimulation terminate afterdischarges caused by cortical stimulation. *Neurology*, 53:2073–2081.

[Longtin, 1991] Longtin, A. (1991). Noise-induced transitions at a Hopf bifurcation in a first-order delay-differential equation. *Phys. Rev. A*, 44:4801–4813.

[Longtin and Hinzer, 1996] Longtin, A. and Hinzer, K. (1996). Encoding with bursting, subthreshold oscillations, and noise in mammalian cold receptors. *Neural Computation*, 8:215–255.

[Longtin et al., 1990] Longtin, A., Milton, J. G., Bos, J. E., and Mackey, M. C. (1990). Noise and critical behavior of the pupil light reflex at oscillation onset. *Phys. Rev. A*, 41:6992–7005.

[Losson et al., 1993] Losson, J., Mackey, M. C., and Longtin, A. (1993). Solution multistability in first-order nonlinear differential delay equations. *Chaos*, 3:167–176.

[Mackey and an der Heiden, 1984] Mackey, M. C. and an der Heiden, U. (1984). The dynamics of recurrent inhibition. *J. Math. Biol.*, 19:211–225.

[Mackey and Glass, 1977] Mackey, M. C. and Glass, L. (1977). Oscillations and chaos in physiological control systems. *Science*, 197:287–289.

[Mackey and Milton, 1987] Mackey, M. C. and Milton, J. G. (1987). Dynamical diseases. *Ann. N. Y. Acad. Sci.*, 504:16–32.

[Mackey and Milton, 1990] Mackey, M. C. and Milton, J. G. (1990). A deterministic approach to survival statistics. *J. Math. Biol.*, 28:33–48.

[Manuca et al., 1998] Manuca, R., Casdagli, M., and Savit, R. (1998). Nonstationarity in epileptic EEG and implications for neural dynamics. *Math. Biosci.*, 147:1–22.

[Matsumoto and Kunisawa, 1978] Matsumoto, G. and Kunisawa, T. (1978). Critical slowing-down near the transition region from the resting to time-ordered states in squid giant axons. *J. Phys. Soc. Japan*, 44:1047–1048.

[Mehta and Schaal, 2002] Mehta, B. and Schaal, S. (2002). Forward models in visuomotor control. *J. Neurophysiology*, 88:942–953.

[Meyer-Lindenberg et al., 2002] Meyer-Lindenberg, A., Ziemann, U., Hajak, G., Cohen, L., and Berman, K. (2002). Transitions between dynamical states of differing stability in the human brain. *Proc Natl Acad Sci USA*, 99:10948–53.

[Miller, 1994] Miller, R. (1994). What is the contribution of axonal conduction delay to temporal structure in brain dynamics? In Pantev, C., editor, *Oscillatory event-related brain dynamics*, pages 53–57, New York. Plenum Press.

[Milton, 2002] Milton, J. (2002). Insights into seizure propagation from axonal conduction times. In Milton, J. and Jung, P., editors, *Epilepsy as a dynamic disease*, pages 15–23, New York. Springer-Verlag.

[Milton and Black, 1995] Milton, J. and Black, D. (1995). Dynamic diseases in neurology and psychiatry. *CHAOS*, 5:8–13.

[Milton and Foss, 1997] Milton, J. and Foss, J. (1997). Oscillations and multistability in delayed feedback control. In Othmer, H. G., Adler, F. R., Lewis, M. A., and Dallon, J. C., editors, *Case studies in mathematical modeling: Ecology, physiology, and cell biology*, pages 179–198, Upper Saddle River, New Jersey. Prentice Hall.

[Milton and Jung, 2002] Milton, J. and Jung, P., editors (2002). *Epilepsy as a dynamic disease*. Springer-Verlag, New York.

[Milton, 1996] Milton, J. G. (1996). *Dynamics of small neural populations*. American Mathematical Society, Providence, Rhode Island.

[Milton, 2000] Milton, J. G. (2000). Epilepsy: multistability in a dynamic disease. In Walleczek, J., editor, *Self-organized biological dynamics and nonlinear control*, pages 374–386, New York. Cambridge University Press.

[Milton et al., 1990] Milton, J. G., an der Heiden, U., Longtin, A., and Mackey, M. C. (1990). Complex dynamics and noise in simple neural networks with delayed mixed feedback. *Biomed. Biochim. Acta*, 49:697–707.

[Milton et al., 1993] Milton, J. G., Chu, P. H., and Cowan, J. D. (1993). Spiral waves in integrate-and-fire neural networks. In Hanson, S. J., Cowan, J. D., and Giles, C. L., editors, *Advances in Neural Information Processing Systems 5*, pages 1001–1007, San Mateo, CA. Morgan Kaufmann.

[Milton et al., 1987] Milton, J. G., Gotman, J., Remillard, G. M., and Andermann, F. (1987). Timing of seizure recurrence in adult epileptic patients: a statistical analysis. *Epilepsia*, 28:471–478.

[Milton et al., 1989] Milton, J. G., Longtin, A., Beuter, A., Mackey, M. C., and Glass, L. (1989). Complex dynamics and bifurcations in neurology. *J. theoret. Biol.*, 138:129–147.

[Milton and Mackey, 2000] Milton, J. G. and Mackey, M. C. (2000). Neural ensemble coding and statistical periodicity: Speculations on the operation of the mind's eye. *J. Physiol. (Paris)*, 94:489–503.

[Morrell, 1985] Morrell, F. (1985). Secondary epileptogenesis in man. *Arch. Neurol.*, 42:318–335.

[Moss et al., 1994] Moss, F., Pierson, D., and O'Gorman, D. (1994). Stochastic resonance: tutorial and update. *Int. J. Bifurc. Chaos*, 4:1383–1397.

[Motamedi et al., 2002] Motamedi, G. K., Lesser, R. P., Miglioretti, D. L., Mizuno-Matsumo, Y., Gordon, B., Webber, W. R. S., Jackson, D. C., Sepkuty, J. P., and Crone, N. E. (2002). Optimizing parameters for terminating cortical afterdischarges with pulse stimulation. *Epilepsia*, 43:836–846.

[Ohira and Yamane, 2000] Ohira, T. and Yamane, T. (2000). Delayed stochastic systems. *Phys. Rev. E*, 61:1247–1257.

[Pakdaman et al., 1998] Pakdaman, K., Grotta-Ragazzo, C., Malta, C. P., Arino, O., and Vibert, J.-F. (1998). Effect of delay on the boundary of the basin of attraction in a system of two neurons. *Neural Networks*, 11:509–519.

[Penfield and Jasper, 1954] Penfield, W. and Jasper, H. (1954). *Epilepsy and the functional anatomy of the human brain*. Churchill, London.

[Platt et al., 1993] Platt, N., Spiegel, E. A., and Tresser, C. (1993). On-off intermittency: A mechanism for bursting. *Phys. Rev. Lett.*, 70:279–282.

[Pomeau and Manneville, 1980] Pomeau, Y. and Manneville, P. (1980). Intermittent transition to turbulence in dissipative systems. *Commun. Math. Phys.*, 74:189–197.

[Rinzel and Baer, 1988] Rinzel, J. and Baer, S. M. (1988). Threshold for repetitive activity for a low stimulus RAMP: A memory effect and its dependence on fluctuations. *Biophys. J.*, 54:551–555.

[S. A. Campbell et al., 1995] S. A. Campbell, S. A., Bélair, J., Ohira, T., and Milton, J. G. (1995). Complex dynamics and multistability in a damped harmonic oscillator with delayed negative feedback. *CHAOS*, 5:640–645.

[Schiff et al., 1994] Schiff, S. J., Jerger, K., Duong, D. H., Chay, T., Spano, M. L., and Ditto, W. L. (1994). Controlling chaos in the brain. *Nature (London)*, 370:615–620.

[Schöner et al., 1986] Schöner, G., Haken, H., and Kelso, J. A. S. (1986). A stochastic theory of phase transitions in human movement. *Biol. Cyner.*, 53:247–257.

[Schwartzkroin and McIntyre, 1997] Schwartzkroin, P. A. and McIntyre, D. C. (1997). Limbic anatomy and physiology. In Engel, Jr., J. and Pedley, T. A., editors, *Epilepsy: A comprehensive textbook*, pages 323–340, Philadelphia. Lippincott-Raven.

[Segev et al., 2002] Segev, R., Benveniste, M., Hulata, E., Cohen, N., Palevski, A., Kapon, E., Shapira, Y., and Ben-Jacob, E. (2002). Long term behavior of lithographically prepared in vitro neuronal networks. *Phys. Rev. Lett.*, 88:118102–1–4.

[Spano et al., 2002] Spano, M. L., Ditto, W. L., Moss, F., and Dolan, K. (2002). Unstable periodic orbits (UPOs) and chaos control in neural systems. In Milton, J. and Jung, P., editors, *Epilepsy as a dynamic disease*, pages 297–322, New York. Springer-Verlag.

[Stanley et al., 1998] Stanley, H. E., Amaral, L. A. N., Andrade, J. S., Buldyrev, S. V., Havlin, S., Makse, H. A., Peng, C.-K., Suki, B., and Viswanathan, G. (1998). Scale-invariant correlations in the biological and social sciences. *Phil. Mag. B*, 77:1373–1388.

[Stark et al., 1958] Stark, L., Campbell, F. W., and Atwood, J. (1958). Pupillary unrest: An example of noise in a biological servo-mechanism. *Nature (London)*, 182:857–858.

[Stépàn, 1989] Stépàn, G. (1989). *Retarded dynamical systems: Stability and characteristic functions*, volume 210 of *Pitman Research Notes in Mathematics Series*. Wiley & Sons, New York.

[Venkataramani et al., 1996] Venkataramani, S. C., Antonsen, T. M., ott, E., and Sommerer, J. C. (1996). On-off intermittency: Power spectra and fractal properties of time series. *Physica D*, 96:66–99.

[Verveen and DeFelice, 1974] Verveen, A. A. and DeFelice, L. J. (1974). Membrane noise. *Prog. Biophys. Mol. Biol.*, 28:253–264.

[Viswanathan et al., 1996] Viswanathan, G. M., Afanasyev, V., Buldyrev, S. V., Murphy, E. J., and Stanley, H. E. (1996). Lévy search patterns of wandering albatrosses. *Nature (London)*, 381:413–415.

[Winfree, 1980] Winfree, A. T. (1980). *The geometry of biological time*. Springer-Verlag, New York.

Chapter 8

ANATOMICAL CONNECTIVITY IN THE CENTRAL NERVOUS SYSTEM REVEALED BY DIFFUSION TENSOR MAGNETIC RESONANCE IMAGING (DT-MRI)

Evren Özarslan
Departments of Physics and Biomedical Engineering
McKnight Brain Institute, University of Florida, Gainesville, FL 32611
evren@mbi.ufl.edu

Thomas H. Mareci
Departments of Biochemistry & Molecular Biology, Physics and Biomedical Engineering
McKnight Brain Institute, University of Florida, Gainesville, FL 32611
thmareci@ufl.edu

Abstract This work explains the basic principles of diffusion weighted and diffusion tensor magnetic resonance imaging (MRI) methods. Both theoretical and experimental aspects are included. Scalar measures derived from diffusion tensors including new anisotropy measures in terms of entropy are presented. Fiber-tract mapping problem is discussed. Limitations of diffusion tensor MRI are included. Directional coherence tensor formalism to smooth the diffusion tensor data is presented. A brief discussion of new approaches to diffusion imaging, such as diffusion spectrum imaging and high angular resolution diffusion imaging, are also discussed. Results from diffusion weighted and diffusion tensor imaging are presented from excised rat spinal cord and brain images acquired at 17.6T.

Keywords: diffusion weighted, diffusion tensor, magnetic resonance imaging, anisotropic diffusion, anisotropy

1. Fundamentals of Diffusion Weighted Magnetic Resonance Imaging

A typical nuclear magnetic resonance experiment starts with the excitation of the nuclei with a 90° radiofrequency (RF) pulse that tilts the magnetization vector into the plane whose normal is along the main magnetic field. Spins that are initially coherent dephase due to many factors most prominent of which are magnetic field inhomogeneities and dipolar interactions. This results in a decay of the electromotive force induced in the receiver. Figure 8.1 shows a spin echo experiment [Hahn, 1950] where a subsequent application of a 180° RF pulse reverses the dephasing due to inhomogeneities and the signal is reproduced. The time between the 90° pulse and reformation of the echo is called TE and it is twice the time between the two RF pulses. This echo is detected by a receiver antenna and is used to produce spectra. Careful application of magnetic field gradients linearly changing in space, enable the acquisition of magnetic resonance images. The details of the techniques employed for spatial encoding will not be explained in this work.

An experiment such as that depicted in Figure 8.1 takes on the order of $10ms$ in a typical imaging experiment. In this timeframe the spins undergo diffusion that results in mixing, which creates attenuation in the signal amplitude when magnetic field inhomogeneities are present. This attenuation can be enhanced by the application of so-called diffusion gradients. Figure 8.2 shows two diffusion gradients added to the previous pulse sequence for this purpose. In this "pulsed gradient spin echo" (PGSE) experiment [Stejskal and Tanner, 1965] two identical gradients around the 180° RF pulse are applied with a time Δ between them. The duration of these gradients are denoted by δ. If $\mathbf{G}$ represents the linear magnetic field gradient applied at time t_1, after the

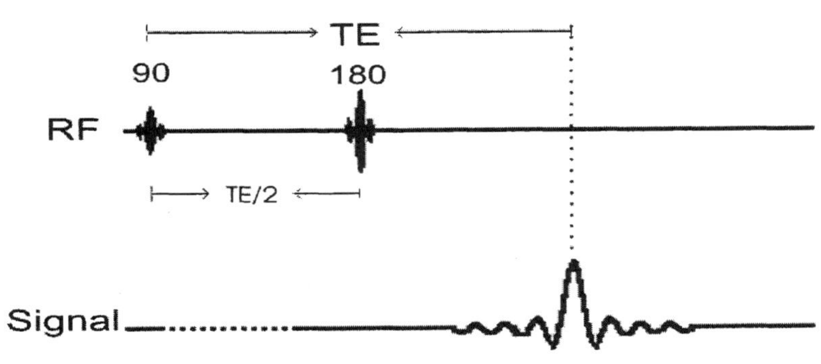

Figure 8.1. Spin echo experiment.

application of the gradient, the spins at the position **r** gain a phase shift

$$\phi_1 = \gamma \int_{t_1}^{t_1+\delta} \mathbf{G} \cdot \mathbf{r}\, dt \, . \tag{1}$$

At a later time t_2, this phase change can be 'undone' by the application of the same gradient along the opposite direction, or alternatively, along the same direction but after the 180° RF pulse. A spin that moves to a different position between the application of these pulses experiences a net phase shift since the phase change due to the subsequent gradient, ϕ_2, will not be equal to $-\phi_1$. As a result, at a particular position in space, there will be many particles with different phases. The signal received from a particular voxel is proportional to the total transverse magnetization within that voxel, given by the two dimensional vector sum of individual magnetizations (each with a magnetic moment μ)

$$M = \sum_{n=1}^{N} \mu e^{i\phi_n} \, . \tag{2}$$

It is clear that when the phases of the spins ϕ_n are all the same, the magnitude of the detected magnetization, M, is equal to its maximum value of $N\mu$; where when the phases are totally random, it is 0. Therefore, the attenuation in the magnitude of the signal indicates the randomness of the phases at a particular position which in turn depends on the randomness in the spins' motional history making it sensitive to the incoherent motion of the molecules alone. Diffusion weighted MRI exploits this phenomenon to quantify diffusion that occurs within the sample.

The dynamics of the magnetization is governed by the phenomenological Bloch Equations [Bloch, 1946]. When diffusion is taken into account, it takes the form [Torrey, 1956]

$$\frac{\partial M_+}{\partial t} = -i\omega_0 M_+ - i\gamma \mathbf{r} \cdot \mathbf{G} M_+ - M_+/T_2 + D\nabla^2 M_+ - \nabla \cdot \mathbf{v} M_+ \, , \tag{3}$$

where $M_+ (= M_x + iM_y)$ is the transverse magnetization, ω_0 is the Larmor frequency, γ is the gyromagnetic ratio, T_2 is the spin-spin relaxation time, D is the diffusion coefficient and last term, quantifying coherent motion of the spins, is included for completeness. From here on, we will call this equation the Bloch-Torrey equation.

For a spin echo experiment, ignoring the last term, Eq. (3) can be solved by using the substitution [Stejskal and Tanner, 1965]

$$M_+(\mathbf{r}, t; \mathbf{G}) = A(t) \exp(-i\omega_0 t - t/T_2 - i\gamma \mathbf{r} \cdot \mathbf{F}) \, , \tag{4}$$

where

$$\mathbf{F}(t) := \int_0^t \mathbf{G}(t')\, dt' - 2\,\Theta\left(t - \frac{TE}{2}\right) \int_0^{\frac{TE}{2}} \mathbf{G}(t')\, dt' \, , \tag{5}$$

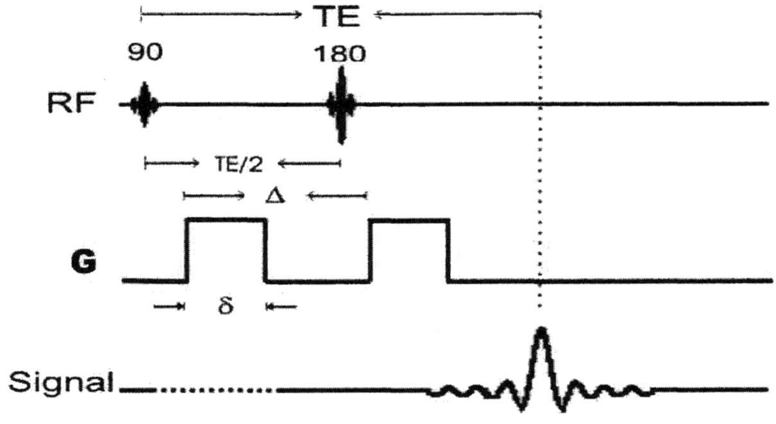

Figure 8.2. Pulsed gradient spin echo experiment. Two diffusion gradients **G** are applied before and after the 180° pulse.

and $\Theta(x)$ is the Heaviside step function which is equal to unity when its argument is positive, and 0 otherwise.

Inserting Eq. (4) into Eq. (3), the resulting first order differential equation yields the Stejskal-Tanner formula for diffusive attenuation [Stejskal and Tanner, 1965]

$$S = S_0 \exp(-\gamma^2 \delta^2 G^2 (\Delta - \delta/3) D) = S_0 \exp(-b\, D) \;, \tag{6}$$

where $S := A(\text{TE})$, $S_0 := A(0)$ and $b := \gamma^2 \delta^2 G^2 (\Delta - \delta/3)$ is the b-factor.

Enhancing diffusive attenuation with the application of gradients introduces a different contrast mechanism in magnetic resonance images. These images are called diffusion weighted images. Moreover, by repeating the experiment with different gradient strengths, it is possible to calculate the diffusion constants. The images where the pixels denote the calculated diffusion coefficients are called quantitative diffusivity maps. Figure 8.3 shows a diffusion weighted image and a quantitative diffusivity map from an excised rat spinal cord.

2. Diffusion Tensor Magnetic Resonance Imaging

Diffusion-weighted images have been utilized extensively in the imaging of neural tissue since it was shown that ischemic strokes can be detected much earlier with diffusion weighted images as compared to traditional T_1 and T_2 weighted images [Moseley et al., 1990].

Neural tissue in central nervous system (CNS) is primarily composed of two distinct areas: white-matter and gray-matter. White-matter includes the axons that transmit signal between different regions. The axons are covered with fatty tissue called the myelin sheath that isolates the intracellular space from

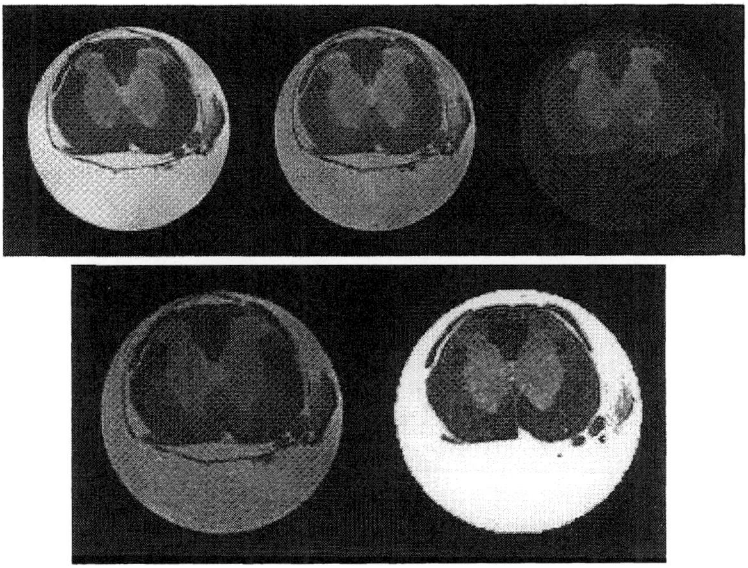

Figure 8.3. Top row includes three diffusion weighted images from an excised rat spinal cord with b-values of approximately 20, 250 and $900s/mm^2$. The bottom left figure is nondiffusion weighted image given by the intercept of the fit. The bottom right image is the quantitative diffusivity map where diffusion sensitizing gradient is along frequency encoding direction (left to right).

extracellular space to increase the speed of signal transmission. Gray-matter is mostly composed of cell bodies and nonmyelinated axons where the axonal architecture is much more complex.

As can be predicted by visually inspecting the electron microscopic images such as those given on pages 52 and 56 of Ref. [Waxman et al., 1995], the diffusion coefficients along different directions in white-matter are different. This diffusional anisotropy is mostly due to the cellular membranes restricting the motion of water molecules, where myelin and the cytoskeleton are contributing factors [Beaulieu and Allen, 1994].

The signal attenuation in diffusion weighted MR images is caused dominantly by the diffusion gradients employed. So, the measured diffusion coefficient depends on the direction along which these gradients are applied. Therefore, by repeating the experiment with gradients applied along many directions, it is possible to quantify the diffusional anisotropy in the voxel. This diffusional anisotropy is presumed to be related to the structural anisotropy in the tissue. The anisotropy maps produced can be expected to give high contrast between the highly oriented areas such as white-matter and other regions.

The fact that diffusion coefficients are anisotropic can be used in yet another way. In highly oriented regions, the diffusion coefficient will be largest along the direction that water will be least restricted. Figure 8.4 shows the effect of sensitizing the signal to diffusional processes along two directions. The top row shows images when the gradients are applied along the direction pointing perpendicular to the page, where the bottom row includes the images when the gradients are applied along up and down direction. When the diffusional process is isotropic (as in free water), the signal attenuation is identical. However, when there is structural anisotropy, the change in the images with increasing b-values depends significantly on the directionality within the tissue. This is most apparent in white-matter. Since water diffusion is least restricted along the fiber directions, the signal attenuates more rapidly when the gradients are applied along those directions as seen in the top row.

Based on this, the direction along which the diffusion coefficient will be greatest can be claimed to give the fiber direction. This is the main hypothesis behind fiber-tract mapping using diffusion-weighted MRI. Starting from a seed point selected within the tissue, repeated stepping in the direction along which diffusion is fastest will allow the mapping of fibers passing through that point [Conturo et al., 1999], [Mori et al., 1999], [Basser et al., 2000], [Özarslan et al., 2001]. If the main hypothesis of fiber-tract mapping is correct, then this idea can be used to map anatomically, hence functionally, connected regions of the brain and spinal cord.

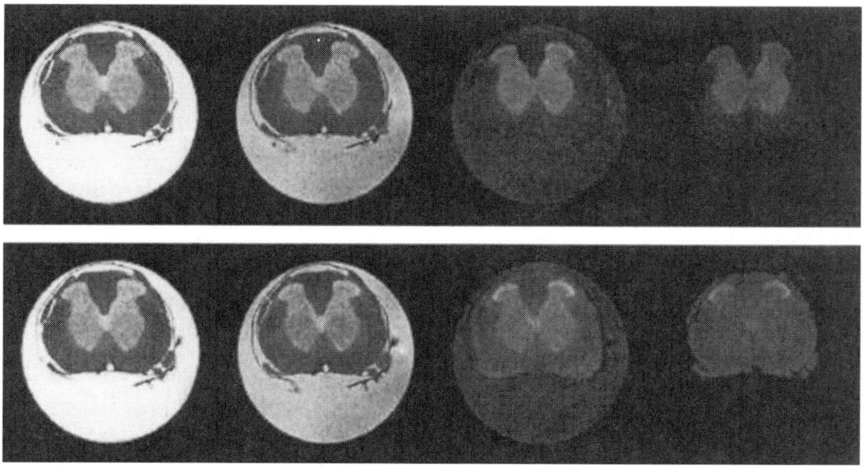

Figure 8.4. Images from an excised spinal cord when the diffusion gradients are applied with increasing strength (from left to right). Top row includes images when the gradients are along the white-matter fibers, where bottom row includes those when the gradients are perpendicular to white-matter fibers.

2.1. Diffusion Tensor Imaging and the Fiber Direction

The most significant quantity needed for fiber-tract mapping is the direction along which the fibers are oriented. This direction has been expected to be that along which diffusion constant will be the greatest. For a general experimentally obtained angular distribution of diffusivities $D(\theta, \phi)$, it can be argued that this direction can be taken as that corresponding to the maximum value of $D(\theta, \phi)$. However, in an experiment that produces discrete samples of this distribution, if only one of the measurements yields an artificially high diffusion constant, then the assumed direction may be wrong. The weakness of taking the direction corresponding to the maximum value of $D(\theta, \phi)$ as the fiber direction is due to its sensitivity to noise. A more appropriate approach would be to fit all samples of the data to a model that accounts for anisotropic diffusion.

Diffusion tensor magnetic resonance imaging (DT-MRI), introduced by Basser et.al. [Basser et al., 1994], employs the anisotropic diffusion model previously proposed by Stejskal [Stejskal, 1965]. In this approach, the diffusion term in Eq.3 was replaced by $\nabla \cdot (\mathbf{D}\nabla M_{+})$, where $\mathbf{D}$ is a second rank (order), real valued, positive definite, symmetric tensor called the diffusion tensor. This change yields the appropriate equation for environments with organizational anisotropy such as liquid crystals [Callaghan, 1991].

Physically, the diffusion coefficient is related to the variance of the displacements in an isotropic environment. This is because, in the unrestricted and isotropic case, solving a diffusion equation or a random walks approach, lead to a molecular displacement profile given by

$$P(\mathbf{x}_0|\mathbf{x}, t) = \frac{1}{(4\pi Dt)^{3/2}} \exp\left(\frac{-|\mathbf{x} - \mathbf{x}_0|^2}{4Dt}\right) . \qquad (7)$$

Here, $P(\mathbf{x}_0|\mathbf{x}, t)$ is the probability that a particle initially at the position $\mathbf{x}_0$ will end up at the position $\mathbf{x}$ at time t. Upon the replacement of the diffusion term with the anisotropic diffusion term, as described above, this displacement probability profile becomes

$$P(\mathbf{x}_0|\mathbf{x}, t) = \frac{1}{((4\pi t)^3 \det(\mathbf{D}))^{1/2}} \exp\left(-\frac{(\mathbf{x} - \mathbf{x}_0)^{\mathrm{T}} \mathbf{D}^{-1} (\mathbf{x} - \mathbf{x}_0)}{4t}\right) . \qquad (8)$$

The components of the diffusion tensor in this equation represents a matrix proportional to the second moments of the displacements. For a more detailed explanation of this, see the appendix.

The full solution to Bloch equation with an anisotropic diffusion term during the PGSE experiment for transverse magnetization is given by

$$\frac{M_+(t)}{M_+(0)} = \exp\Bigg(- i\omega_0 t - \tfrac{t}{T_2} - i\gamma \mathbf{r} \cdot \left[\mathbf{F}(t) - 2\Theta\left(t - \tfrac{TE}{2}\right) \mathbf{F}\left(\tfrac{TE}{2}\right)\right]$$
$$- \gamma^2 \Bigg[\textstyle\int_0^t \mathbf{F}^T \mathbf{D} \mathbf{F} dt' - 4\Theta\left(t - \tfrac{TE}{2}\right) \mathbf{F}\left(\tfrac{TE}{2}\right)^T$$
$$\mathbf{D} \left(\textstyle\int_{TE/2}^t \mathbf{F}(t') dt' - \mathbf{F}\left(\tfrac{TE}{2}\right)\left(t - \tfrac{TE}{2}\right)\right) \Bigg] \Bigg) \tag{9}$$

where Θ is the Heaviside step function, and here $\mathbf{F}(t)$ is defined to be

$$\mathbf{F}(t) := \int_0^t \mathbf{G}(t')\, dt' \,. \tag{10}$$

Note that the last term in Eq.9 quantifies the diffusive attenuation and will be our focus from here on. With the definitions $S = |M_+(TE)|$ and $S_0 = |M_+(0)|\, e^{-TE/T_2}$, we can easily see that

$$\frac{S}{S_0} = \exp\left(-\gamma^2 \int_0^{TE} \boldsymbol{\mathcal{F}}(t')^T\, \mathbf{D}\, \boldsymbol{\mathcal{F}}(t')\, dt' \right) , \tag{11}$$

where

$$\boldsymbol{\mathcal{F}}(t) := \mathbf{F}(t) - 2\,\Theta\left(t - \frac{TE}{2}\right) \mathbf{F}\left(\frac{TE}{2}\right) . \tag{12}$$

In DT-MRI a b-matrix, defined by [Basser et al., 1994]

$$\mathbf{b} := \gamma^2 \int_0^{TE} \boldsymbol{\mathcal{F}}(t) \boldsymbol{\mathcal{F}}(t)^T\, dt \,, \tag{13}$$

is used to calculate an effective diffusion tensor through the equation

$$\frac{S}{S_0} = e^{-\mathrm{trace}(\mathbf{b}\mathbf{D}^{\mathrm{eff}})} , \tag{14}$$

where it is assumed that diffusion tensor is almost constant in time so that $\mathbf{D}^{\mathrm{eff}} \simeq \mathbf{D}$.

This equation has seven unknowns where one of them is the signal intensity if there were no diffusion (S_0) and the remaining six are the unique elements of the diffusion tensor. Therefore, seven experiments producing seven linearly independent equations are sufficient to determine the diffusion tensor. Note that Eq.14 is the corresponding expression to that in Eq.6 that enables the calculation of diffusion coefficients. Figure 8.5 shows the components of the diffusion tensor of an excised rat brain.

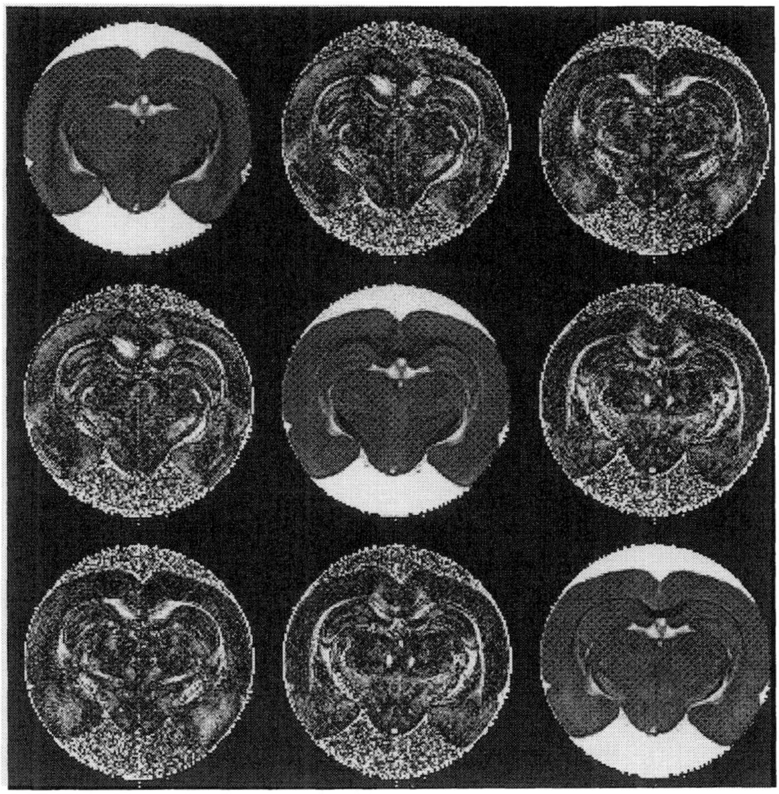

Figure 8.5. Diffusion tensor calculated from a series of diffusion weighted images. The order of the images is $D_{xx}, D_{xy}, D_{xz}, D_{yx}, D_{yy}, D_{yz}, D_{zx}, D_{zy}, D_{zz}$ from left to right and top to bottom. The absolute values of the offdiagonal elements are taken since they can be negative valued.

So we have two experimental schemes. In one of them, we can apply diffusion gradients along certain direction $\hat{\mathbf{g}}$ with differing gradient strengths to calculate the apparent diffusion coefficient $D(\hat{\mathbf{g}})$ from Eq.6. Here $\hat{\mathbf{g}}$ is a unit vector representing the direction of the diffusion gradient, i.e., $\mathbf{G} = G\hat{\mathbf{g}}$. In the second scheme, we can apply the diffusion gradient in at least six noncoplanar directions and with differing gradient strengths and calculate the apparent diffusion tensor $(\mathbf{D})$ by using Eq.14. An important question to be addressed is how these results are related. If the background (imaging) gradients are small compared with the diffusion gradients, which is almost always the case, then comparison of Eq. 6 with Eq. 14 yields the simple relation [Hsu and Mori, 1995]

$$D(\hat{\mathbf{g}}) = \hat{\mathbf{g}}^T \mathbf{D} \, \hat{\mathbf{g}} \, . \tag{15}$$

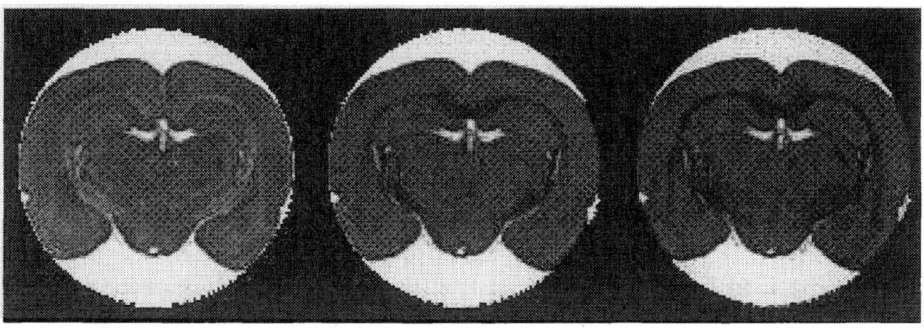

Figure 8.6. Eigenvalues of the diffusion tensor in decreasing order.

In addition to its great reduction in the number of experiments required to quantify three dimensional distribution of diffusivities, DT-MRI also makes it possible to quantify the direction along which diffusion coefficient is greatest. This is done by finding $\hat{\mathbf{g}}$ that maximizes D. Once the diffusion tensor is calculated, this direction is just the eigenvector corresponding to the largest of the eigenvalues of $\mathbf{D}$.

3. Scalar Measures Derived From DT-MRI

Similar to what was presented for the case of diffusion weighted imaging, fitting the data to the Stejskal-Tanner relation for anisotropic diffusion, Eq.14, produces an almost nondiffusion weighted image which is calculated from the intercept of the fit. The intercept does not have any diffusion weighting due to the gradients used in the pulse sequence since it corresponds to the signal at $b = 0$. Therefore, the only diffusion weighting comes from the magnetic field inhomogeneities within the sample. This reduction in the diffusion weighting in the MR images can only be obtained by the quantification of diffusion.

A diffusion tensor has 9 components. However since it is a symmetric tensor, 6 of these have distinct values. Also because it is symmetric, diffusion tensor is diagonalizable. Its spectral decomposition is given by

$$\mathbf{D} = \lambda_1\, \mathbf{u}_1\mathbf{u}_1^{\mathrm{T}} + \lambda_2\, \mathbf{u}_2\mathbf{u}_2^{\mathrm{T}} + \lambda_3\, \mathbf{u}_3\mathbf{u}_3^{\mathrm{T}}\,, \tag{16}$$

where λ_i are the eigenvalues and $\mathbf{u}_i$ are the eigenvectors. Images of the three eigenvalues of the diffusion tensor calculated on an excised rat brain is given in Figure 8.6. Upon rotations, the eigenvalues will remain constant where the components of the eigenvectors will change. Therefore, an (rotationally) invariant basis of the diffusion tensor has three elements. These three elements can be chosen to be the three eigenvalues. However, it is not easy to interpret the meaning of these values perhaps except the principal (greatest) eigenvalue, which is just the diffusivity along the fiber direction. Another choice could

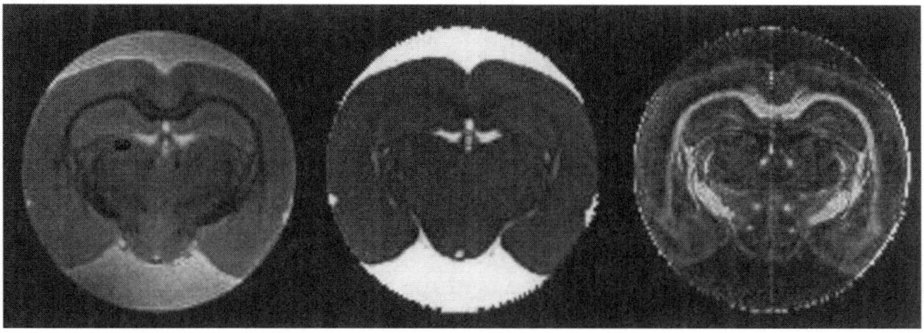

Figure 8.7. Nondiffusion weighted image (left), mean diffusivity (middle), and fractional anisotropy (right) images from the excised rat brain.

be the first three moments of the diffusion tensor, namely trace($\mathbf{D}$), trace($\mathbf{D}^2$) and trace($\mathbf{D}^3$). All of these indices, like the eigenvalues, have units. The first moment of the diffusion tensor, giving the average of diffusivities along all directions, has been chosen to be the first scalar measure derived from the diffusion tensor. This quantity, called mean diffusivity [Basser, 1995] is given by

$$\langle D \rangle = \frac{\text{trace}(\mathbf{D})}{3} = \frac{\lambda_1 + \lambda_2 + \lambda_3}{3} \,. \tag{17}$$

Traditionally, it has been found useful to have unitless scalars instead of the second and third moments. Especially an index that measures the level of organization in a pixel would be very useful since it would make it easy to distinguish highly organized regions, like white-matter, from others. There have been several suggestions for the parametrization of anisotropy in the tissue. The most commonly used one is Fractional Anisotropy [Basser, 1995], which is expressed in terms of the eigenvalues as

$$FA = \sqrt{\frac{3}{2}}\sqrt{1 - \frac{(\lambda_1 + \lambda_2 + \lambda_3)^2}{3(\lambda_1^2 + \lambda_2^2 + \lambda_3^2)}} \,. \tag{18}$$

FA takes the value of 0 for totally isotropic tensors whereas it takes the value of 1 in the complete anisotropic case. As for the third invariant, a scalar measure such as skewness [Bahn, 1999] can be used. However, in neural tissue the third invariants have not found widespread utilization. In Figure 8.7 we show the nondiffusion weighted image of the rat brain calculated from the intercept of the fit along with mean diffusivity and fractional anisotropy maps.

3.1. Anisotropy in terms of Entropy

An important matter in fiber-tract mapping is choosing an appropriate criterion to terminate the tract tracing process. Traditionally, this criterion has been the fall of anisotropy below a prespecified value. The idea behind this is that in low anisotropy regions, the fiber direction becomes too uncertain to continue tracking. This idea in effect states what is expected of an anisotropy measure. Namely, anisotropy should quantify the level of orientational certainty contained in the diffusion data. Or equivalently, isotropy can be defined as the level of uncertainty. With this in mind, we take the orientation specified by the solid angle Ω (corresponding to the polar angle θ and azimuthal angle ϕ) as a random variable and define the isotropy in terms of the differential entropy given by

$$\sigma := -\frac{3}{2\pi}\int_0^{2\pi} d\phi \int_0^{\pi/2} d\theta \sin\theta D_N(\theta,\phi) \ln D_N(\theta,\phi) \,, \qquad (19)$$

where D_N is the 'normalized' diffusivity given by

$$D_N(\theta,\phi) := D(\theta,\phi) \Big/ \frac{3}{2\pi}\int_0^{2\pi} d\phi \int_0^{\pi/2} d\theta \sin\theta D(\theta,\phi) \,. \qquad (20)$$

Integrations are carried out over the hemisphere because diffusivities have antipodal symmetry, therefore directions correspond to rays rather than vectors. Eq.19 is the same expression with the Boltzmann-Gibbs entropy, which quantifies the uncertainty in a phase space taken here to be the projective 2-sphere and $D_N(\theta,\phi)$ is the probability function.

3.2. Anisotropy of a Diffusion Tensor

Having determined a measure of certainty in a given diffusivity profile, we now find a corresponding relation to Eq.19 from the diffusion tensor. This can be done by discretizing $D(\hat{\mathbf{g}})$ in Eq.15 and using discretized versions of Eqs. 19 and 20. However, this would require too many computations, and the differential entropy has the undesirable feature that the entropy calculated from samples of a continuous distribution does not converge asymptotically to the entropy calculated from the continuous distribution. Therefore, we look for an expression that is calculated directly from the tensor. To do this, we realize that the trace of a 3×3 matrix $\mathbf{A}$ is given by

$$\mathrm{trace}(\mathbf{A}) = \frac{3}{2\pi}\int_0^{2\pi} d\phi \int_0^{\pi/2} d\theta \sin\theta \, \boldsymbol{\psi}_{\theta,\phi}^T \mathbf{A} \, \boldsymbol{\psi}_{\theta,\phi} \,, \qquad (21)$$

where $\boldsymbol{\psi}_{\theta,\phi}^T = (\sin\theta \, \cos\phi, \, \sin\theta \, \sin\phi, \, \cos\theta)$. We note that this is the same expression as the denominator in Eq.20 when $D(\theta,\phi)$ is replaced by the quadratic form in Eq.15. So, Eq.20 turns into

$$D_N(\theta,\phi) = \frac{\boldsymbol{\psi}_{\theta,\phi}^T \mathbf{D} \, \boldsymbol{\psi}_{\theta,\phi}}{\mathrm{trace}(\mathbf{D})} = \boldsymbol{\psi}_{\theta,\phi}^T \boldsymbol{\rho} \, \boldsymbol{\psi}_{\theta,\phi} \,, \qquad (22)$$

where

$$\boldsymbol{\rho} := \frac{\mathbf{D}}{\mathrm{trace}(\mathbf{D})} \,. \tag{23}$$

Note that $\boldsymbol{\rho}$ is a symmetric positive definite matrix with $\mathrm{trace}(\boldsymbol{\rho}) = 1$. Inserting Eq. 22 into Eq.19, we get

$$\begin{aligned} \sigma &= -\frac{3}{2\pi}\int_0^{2\pi} d\phi \int_0^{\pi/2} d\theta \sin\theta \, \boldsymbol{\psi}_{\theta,\phi}^T \boldsymbol{\rho}\, \boldsymbol{\psi}_{\theta,\phi} \ln\left(\boldsymbol{\psi}_{\theta,\phi}^T \boldsymbol{\rho}\, \boldsymbol{\psi}_{\theta,\phi}\right) \\ &\simeq -\frac{3}{2\pi}\int_0^{2\pi} d\phi \int_0^{\pi/2} d\theta \sin\theta \, \boldsymbol{\psi}_{\theta,\phi}^T \boldsymbol{\rho} \ln \boldsymbol{\rho}\, \boldsymbol{\psi}_{\theta,\phi} \,, \end{aligned} \tag{24}$$

where the approximation used is the same as the classical approximation to quantum entropy [Wehrl, 1978].

Using Eq.21 once more, we get the simple expression for the entropy

$$\sigma(\boldsymbol{\rho}) := -\mathrm{trace}(\boldsymbol{\rho} \ln \boldsymbol{\rho}) = -\sum_{i=1}^{3} \rho_i \ln \rho_i \simeq \sigma \,, \tag{25}$$

where ρ_i are the eigenvalues of $\boldsymbol{\rho}$. Note that this is the same expression as von Neumann entropy in quantum statistical mechanics (and quantum information theory) quantifying the uncertainty level (or the amount of lack of information) contained in a density matrix [Fick and Sauermann, 1990].

Since isotropy is a measure of uncertainty, we define anisotropy to be a function monotonically decreasing with increasing entropy. Two such indices that we propose are: quantitative anisotropy that quantifies the orientational information content of a diffusion tensor given by [Özarslan and Mareci, 2003a]

$$\mathrm{QA} = \ln 3 - \sigma \,, \tag{26}$$

and visual anisotropy that gives better contrast in calculated images given by

$$\mathrm{VA} := \sqrt{2 + \frac{1}{\frac{\sigma}{2\ln 3} - 1}} \,. \tag{27}$$

The formulations of the anisotropy indices QA and VA as in Eqs. 26 and 27 enables one to visualize the behavior of these indices for given values of eigenvalues of $\boldsymbol{\rho}$. If we order the eigenvalues such that $\rho_1 \geq \rho_2 \geq \rho_3$, then this condition along with $\rho_1 + \rho_2 + \rho_3 = 1$ and the positive definiteness of the eigenvalues, limits the allowed values of ρ_1 and ρ_2 to the interior of a triangular zone in the $\rho_1\rho_2$ plane given by:

$$\begin{aligned} \rho_1 - \rho_2 &\geq 0 \\ \rho_1 + 2\rho_2 &\geq 1 \\ \rho_1 + \rho_2 &\leq 1 \,. \end{aligned} \tag{28}$$

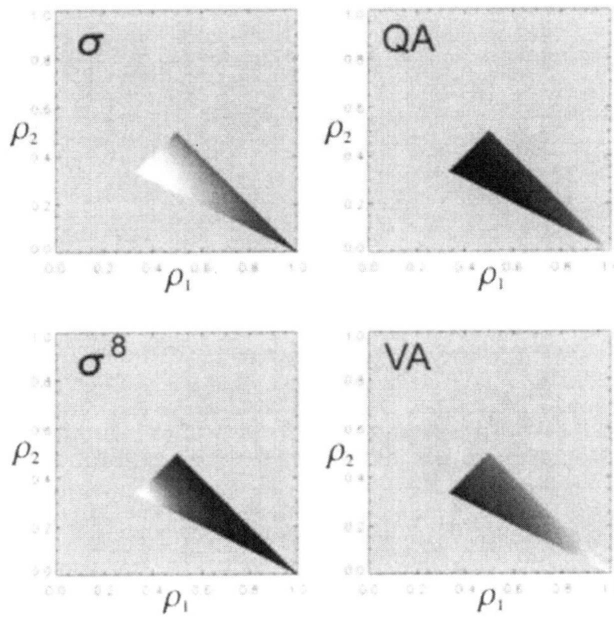

Figure 8.8. The behavior of entropy and anisotropy indices for all possible eigenvalues of the normalized diffusion tensor.

Since ρ_3 is uniquely defined by a given set of ρ_1 and ρ_2, these anisotropy measures can be visualized within this triangle. Figure 8.8 shows these maps for these two indices along with von Neumann entropy and its 8th power.

Figure 8.9 shows the calculated maps of these indices for a coronal rat brain slice. As expected, regions corresponding to white-matter are bright in anisotropy maps, and dark in entropy maps making them a suitable choice as a thresholding parameter in fiber-tracking.

The inflation in the number of anisotropy indices that have been proposed in the literature is a serious problem because given two voxels, one index may identify the first one as more anisotropic, whereas another may identify it as more isotropic. This makes the comparative analyses of the anisotropy values dependent on the choice of the anisotropy index. The anisotropy indices proposed previously lack the meaning that we have given to it, namely, a quantity increasing with increasing information content. Moreover, FA index that we have mentioned above can be expressed in terms of the normalized diffusion tensor $\boldsymbol{\rho}$ we defined:

$$FA = \sqrt{\frac{1}{2}\left(3 - \frac{1}{\text{trace}(\boldsymbol{\rho}^2)}\right)}. \tag{29}$$

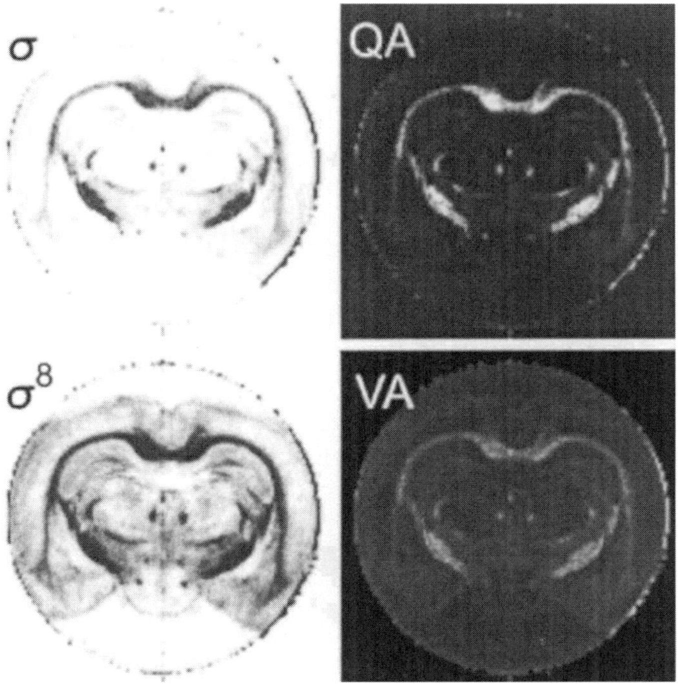

Figure 8.9. Images formed by calculating the scalar indices introduced in the text for a transverse excised rat brain slice. Top left is the entropy, bottom left is its 8th power, top right is QA, and bottom right is VA.

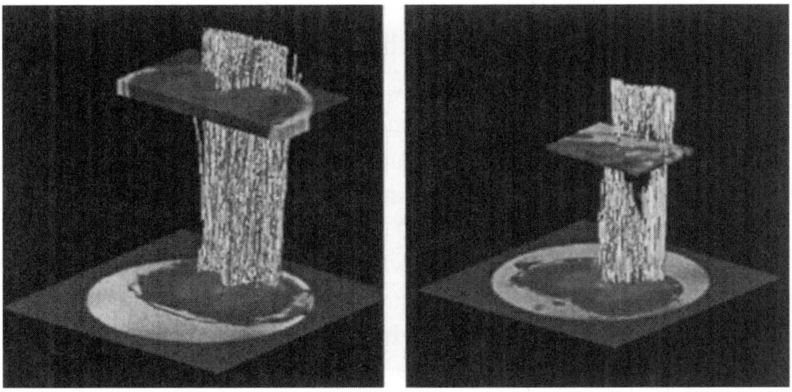

Figure 8.10. White matter fiber tracts of normal (left) and injured (right) rat spinal cords.

So, just like VA is a scaled version of trace(ρ ln ρ), FA is a scaled version of trace(ρ^2). This is just the purity index in the information theory and statistical mechanics literature and is known to be a good measure to decide whether or not the distribution is pure, while it violates the properties that a valid measure of information should possess [Wehrl, 1978] [Fick and Sauermann, 1990].

4. Fiber-Tract Mapping in Neural Tissue

Once the diffusion tensor is calculated, the scalar index that will be used as the termination criterion, as well as the eigenvalues and eigenvectors, are calculated. The fiber tracts are constructed by repeatedly stepping in the direction along which the diffusion coefficient is greatest, which is given by the eigenvector corresponding to the largest of the eigenvalues. The constructed fiber-tracts can be thought of as curves whose tangent vector is set equal to this direction. The equation of motion in this case is given by a Frenet formula [Basser et al., 2000]:

$$\frac{d\mathbf{r}}{ds} = \epsilon_{>} \,, \tag{30}$$

where $\mathbf{r}$ is the position vector, ds is the infinitesimal scalar distance on the curve, and $\epsilon_{>}$ is the unit vector from diffusion data. This equation can be solved by integration with user supplied initial condition.

Excised spinal cords of one normal and one injured female Sprague-Dawley rats were imaged with a diffusion weighted spin-echo pulse sequence at 17.6T using a Bruker Avance imaging and spectrometer system. Diffusion gradients were applied along seven directions: $\mathbf{e}_x$, $\mathbf{e}_y$, $\mathbf{e}_z$, $\frac{1}{\sqrt{2}}(\mathbf{e}_x + \mathbf{e}_y)$, $\frac{1}{\sqrt{2}}(\mathbf{e}_x + \mathbf{e}_z)$, $\frac{1}{\sqrt{2}}(\mathbf{e}_y + \mathbf{e}_z)$, $\frac{1}{\sqrt{3}}(\mathbf{e}_x + \mathbf{e}_y + \mathbf{e}_z)$ with four different strengths 0, 180, 360 and $540mT/m$. Imaging parameters were: TR $= 1000ms$, TE $= 27.1ms$, $\Delta = 17.8ms$, $\delta = 2.4ms$, resolution$= 40 \times 40 \times 250\mu m^3$. Fibers were calculated from selected regions of interest (ROIs) which specify the initial values of Eq. 30. Tracking is terminated when anisotropy drops below a certain prespecified value. Figure 8.10 shows an example of calculated fibers from normal and injured spinal cords when the ROIs are selected in the dorsal column. The injury site suffers a discontinuity in the fibers.

Also imaged was an excised rat brain with b-values of approximately 100, 500, $1000s/mm^2$. Imaging parameters were: TR $= 3060ms$, TE $= 28.8ms$, $\Delta = 17.8ms$, $\delta = 2.4ms$, resolution$= 117 \times 117 \times 270\mu m^3$, and matrix size was $128 \times 128 \times 78$. The seed points were selected in corpus callosum and cerebral peduncles and fibers were tracked in orthograde as well as retrograde directions. These images are shown in Figure 8.11.

5. Problems of DT-MRI Based Fiber Tracking

Despite the fact that only limited quantitative verification of the DT-MRI based fiber-tract mapping results has been done, qualitative comparisons with known anatomy of the nervous tissue has shown that DT-MRI is able to correctly map the major axonal pathways [Özarslan et al., 2001]. Despite the promising results achieved by DT-MRI, it is known that it has significant weaknesses.

One of the problems with DT-MRI is that the formalism presented to calculate the diffusion tensor from signal attenuation does not account for spatial dependence of the diffusion tensor within a voxel. It is the correct description for a medium that is homogeneously anisotropic such as liquid crystals. It is not known how spatial dependence of the tensor would affect the measured signal attenuation.

Another problem with the formalism just presented, is that it assumes no boundary conditions, hence assumes free diffusion and does not account for restricting boundaries. These problems are present in diffusion weighted imaging as well. Therefore, the word 'apparent' has been used before the phrases 'diffusion coefficient' and 'diffusion tensor' to emphasize that calculated diffusivities are in reality dependent on the parameters of the pulse sequence used.

DT-MRI based fiber tracking assumes that there is a single fiber direction within a voxel. A typical axon has a diameter in the order of $10\mu m$ and typical voxel volume of the images we acquire at our institution is in the order of $(100\mu m)^3$. So, in every voxel, there is a bundle of axons. It is known that in many areas of the brain, these axons cross. This may result in deviations in the calculated directions and premature termination of tracking due to a decrease

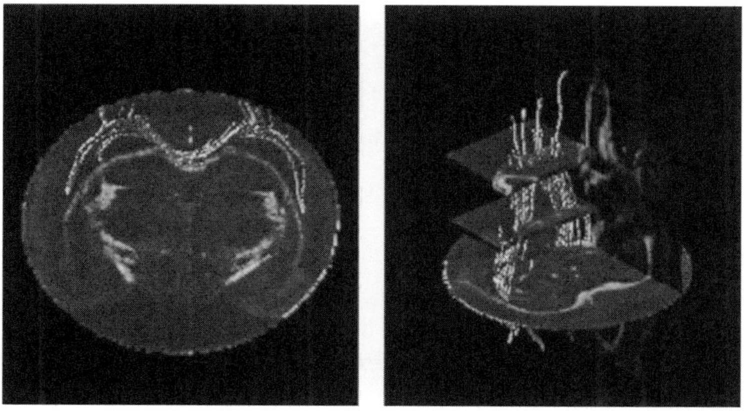

Figure 8.11. White matter fiber tracts in the corpus callosum (left) and cerebral peduncles (right) in a rat brain.

in the anisotropy value. This effect will be more serious in clinical imaging, where voxel volume will be in the order of $1mm^3$. Similar problems may occur even when there is a single fiber bundle in the voxel with a curvature in the order of $1/d$ or higher, where d is the length of the edge of a voxel.

5.1. A Post Processing Algorithm for Diffusion Tensor Images

The relatively less demanding nature of DT-MRI, has directed some groups to overcome the difficulties by designing post processing algorithms that improve the fiber tract mapping process. As an example we present our directional coherence formalism [Özarslan and Mareci, 2002].

Diffusion-weighted imaging based fiber-tract mapping has been thought of as a technique to be applied to white-matter because of the high certainty in the fiber direction. However, we have managed to map structure in spinal cord gray-matter by two means, first by defining a scalar index called directional coherence (DC) that quantifies how consistent DT-MRI based fiber direction in a voxel is with its neighboring voxels. Second, we smooth the diffusion tensor image in such a way that the smoothed tensor field will have higher orientational certainty. Diagonalization of the new tensor yields more robust tracking making it possible to see the structure in gray-matter.

Calculation of DC proceeds as follows. First diffusion tensors are diagonalized and the primary eigendirection is called $\hat{\mathbf{u}}$. Then a matrix T is defined to be the outer product of this vector with itself, i.e. $\mathbf{T}_0 := \hat{\mathbf{u}}\hat{\mathbf{u}}^T$. Next, this tensor image is convolved with a Gaussian kernel of standard deviation s given by

$$K_s(\mathbf{r}) = \frac{1}{(2\pi s^2)^{3/2}} e^{-\frac{-|\mathbf{r}|^2}{2s^2}} \tag{31}$$

to give

$$\mathbf{T}_s(\mathbf{r}) := (K_s * \mathbf{T}_0)(\mathbf{r}). \tag{32}$$

Then directional coherence is defined by

$$\mathrm{DC}_s(\mathbf{D}) := \mathrm{VA}(\mathbf{T}_s). \tag{33}$$

Obviously if the voxel is in a perfectly coherent environment, the convolution step will not change the matrix, i.e. T_s will remain equal to T_0 which has the highest possible value of 1 for VA. Therefore, starting with a unidirectional matrix, Gaussian convolution 'contaminates' its purity by its environment, and reduced anisotropy is quantified as the level of coherence in the local environment.

Next, we smooth the diffusion tensor field D based on the idea that given two voxels, the 'interacting' quantity between the diffusion tensors from each voxel should be their quadratic form along the direction connecting the two voxels.

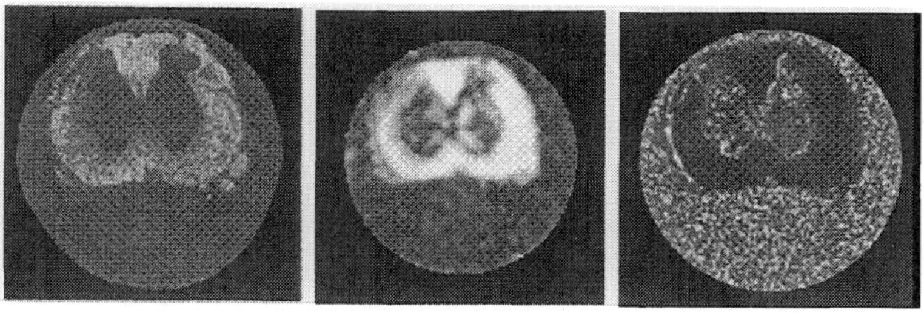

Figure 8.12. On the left is the visual anisotropy of the diffusion tensor VA(**D**), next to it is VA($\mathbf{T}_{\mathrm{DC}}^{.5,1.0}$) which is a measure of how coherent the fibers in a particular voxel with its neighbours is. Finally on the right is the angle by which the principal eigenvector of the diffusion tensor is changed as a result of the smoothing algorithm.

A simple implementation of this idea is to define a connectivity vector by the equation

$$\mathbf{C}_{\mathrm{ijk}} = \frac{1}{2}\begin{pmatrix} \pm\hat{\mathbf{x}}^T(D^s_{\mathrm{i+1jk}} + D^s_{\mathrm{i-1jk}})\,\hat{\mathbf{x}} \\ \pm\hat{\mathbf{y}}^T(D^s_{\mathrm{ij+1k}} + D^s_{\mathrm{ij-1k}})\,\hat{\mathbf{y}} \\ \pm\hat{\mathbf{z}}^T(D^s_{\mathrm{ijk+1}} + D^s_{\mathrm{ijk-1}})\,\hat{\mathbf{z}} \end{pmatrix}, \tag{34}$$

where $\pm$ signs in front of each of the components indicate the uncertainty in the sign at this point. Next, we calculate the outer product of this vector with itself and do a Gaussian convolution and define the directional coherence tensor as

$$(\mathbf{T}^{s,t}_{\mathrm{DC}})_{\mathrm{ijk}} = K_t * \mathbf{C}^{'}_{\mathrm{ijk}}\mathbf{C}^{'\,T}_{\mathrm{ijk}}, \tag{35}$$

where $\mathbf{C}^{'}_{\mathrm{ijk}}$ is the same with $\mathbf{C}_{\mathrm{ijk}}$ with the combination of signs that maximizes $\mathrm{trace}(\mathbf{T}^{s,t}_{\mathrm{DC}}\,\mathbf{D})$. Note that $\mathrm{VA}(\mathbf{T}^{s,t}_{\mathrm{DC}})$ is the directional coherence of the new tensor field.

Figure 8.12 shows the scalar indices that are produced from the algorithm. The figure on the left is just the Visual Anisotropy image of a slice from the excised rat spinal cord. The image in the middle is the VA of the directional coherence tensor from this slice, which is the coherence in the smoothed tensor field. As it is obvious from the comparison, unlike the VA of the diffusion tensor this image contains high contrast within the gray-matter. Although the diffusion tensors in these bright pixels in the gray-matter are quite isotropic, they are coherent directionally. The image on the right depicts the angle by which the presumed fiber directions are changed as a result of the algorithm. It is clear that the algorithm does not affect the highly anisotropic and highly coherent areas in white-matter, whereas there are changes in gray matter. In nearly isotropic areas, the directions when chosen to be the eigenvector corresponding to the largest of the eigenvalues, have the danger to be false as much as 90° due to

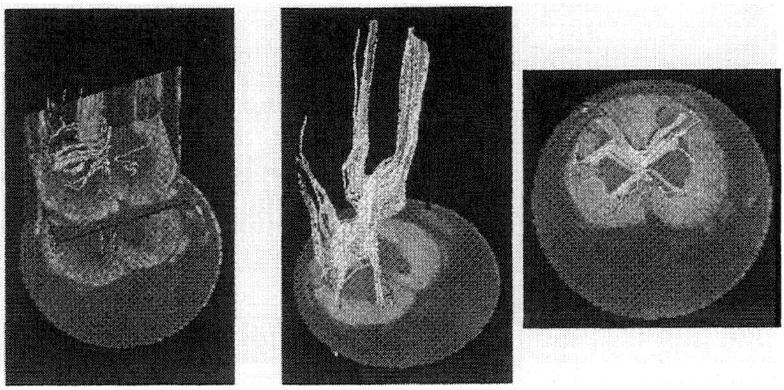

Figure 8.13. These images show the change in the calculated fiber-tracts. The first of these show the erratic trajectories in gray-matter before smoothing, and last two show the calculated fiber-tracts from the smoothed tensor image.

the bias in the sorting of the eigenvalues. The very bright pixels in gray-matter corresponds to the correction due to this kind of errors.

The fiber-tracts are calculated for the excised rat spinal cord as can be seen in Figure 8.13. The left figure shows the erratic trajectories when the ROI is chosen in gray-matter. The figures in the middle and on the right show the calculated fibers after the algorithm is applied. The structure in gray-matter is significantly improved. These results are consistent with the directional coherence map because the results of the fiber-tract mapping algorithm show coherent fiber bundles in the areas where directional coherence values were high. And the dark area in the center correspond to fiber crossings in the calculated fiber-tracts.

5.2. New Approaches to Diffusion Imaging

Recently, there has been attempts to replace the tensor model with more sophisticated models. According to one approach [Wedeen et al., 2000], q-space imaging methods (see appendix for a brief discussion) have been applied to calculate displacement profiles of molecules without satisfying the requirement for the narrow pulse approximation assumed in the q-space formalism [Callaghan, 1991]. This method called 'diffusion spectrum imaging' is unlikely to be used clinically because of very demanding hardware and acquisition times necessary [Basser, 2002].

Another approach to diffusion imaging is called 'high angular resolution diffusion imaging' [Tuch et al., 1999]. In this method, the same gradient strength is applied along many directions to calculate a diffusivity profile $D(\hat{\mathbf{g}})$ modeling three dimensional diffusion by a series of one dimensional diffusion processes.

This approach was further improved, where the spherical harmonic transform of this diffusivity profile is taken, and the resulting Laplace series is terminated so that it will include only the significant terms in the expansion [Frank, 2002].

We have further improved this approach by writing a corresponding Bloch-Torrey equation that includes a Cartesian tensor of rank higher than 2 [Özarslan and Mareci, 2003b]. In this case, the corresponding Stejskal-Tanner relationship is derived allowing the evaluation of the components of higher rank tensors by using least squares fits. This makes the evaluation of the spherical harmonic transform unnecessary. The calculation of anisotropy and fiber directions necessary for fiber-tract mapping remains an open area of research for these new techniques.

Acknowledgments

The authors would like to thank Dr. S. Michelle DeFord and Dr. Ronald L. Hayes for providing the rat brain. The magnetic resonance images were acquired in the Advanced Magnetic Resonance Imaging and Spectroscopy Facility of the McKnight Brain Institute.

Appendix: Diffusion Tensor and Displacement Profile

In this appendix, we try to develop an intuitive understanding of what diffusion tensor imaging really performs. To do this, we give a brief review of Markoff's method applied to the problem of random flights and relate it to another MR diffusion imaging modality called q-space imaging.

Assume that we have a single particle undergoing a series of random flights [Chandrasekhar, 1943] where at each step, the particle moves a distance $\mathbf{r}_k = (x_k, y_k, z_k)^T$ with a probability distribution function $\tau_k(x_k, y_k, z_k)$. Then the probability of finding the particle at the position $\mathbf{R} = \sum_{k=1}^{N} \mathbf{r}_k$ relative to the initial position is given by

$$W_N(\mathbf{R}) = \frac{1}{8\pi^3} \int_{-\infty}^{\infty} \exp(-i\boldsymbol{\lambda} \cdot \mathbf{R})\, A_N(\boldsymbol{\lambda})\, d\boldsymbol{\lambda}\ , \tag{8.A.1}$$

where

$$A_N(\boldsymbol{\lambda}) = \prod_{k=1}^{N} \int_{-\infty}^{\infty} \tau_k(\mathbf{r}_k)\, \exp(i\boldsymbol{\lambda} \cdot \mathbf{r}_k)\, d\mathbf{r}_k\ . \tag{8.A.2}$$

We will be interested in the bahaviour of this general solution under two conditions: $N \to \infty$ and probability function $\boldsymbol{\tau}_k$ is identical at each step, i.e., $\boldsymbol{\tau} = \boldsymbol{\tau}_k$ for all k. If we further assume that all first moments of displacements vanish, we get the expression:

$$A_N(\boldsymbol{\lambda}) = \exp\left(-\frac{1}{2}\boldsymbol{\lambda}^T \mathbf{S} \boldsymbol{\lambda}\right)\ , \tag{8.A.3}$$

where

$$\mathbf{S} := \begin{pmatrix} \langle x^2 \rangle & \langle xy \rangle & \langle xz \rangle \\ \langle yx \rangle & \langle y^2 \rangle & \langle yz \rangle \\ \langle zx \rangle & \langle zy \rangle & \langle z^2 \rangle \end{pmatrix} , \tag{8.A.4}$$

and components of this matrix are given by

$$S_{ij} = \langle r_i r_j \rangle = N \int d\mathbf{r} \; r_i \, r_j \, \tau(\mathbf{r}) \, . \tag{8.A.5}$$

This yields a probability distribution

$$W_N(\mathbf{R}) = \left(8\pi^3 \det(\mathbf{S})\right)^{-\frac{1}{2}} \exp\left(-\frac{1}{2}\mathbf{R}^T \mathbf{S}^{-1}\mathbf{R}\right) . \tag{8.A.6}$$

Now we look at what happens to the signal when the gradient pulses are much shorter than the interval between them, i.e., $\delta \ll \Delta$ in a PGSE experiment. Under this condition, all the diffusive processes occur in the absence of diffusion gradients. Therefore, it is possible to quantify the signal attenuation by the relation [Callaghan, 1991]

$$\frac{S(\mathbf{g})}{S_0} = \int d\mathbf{r} \; \rho(\mathbf{r}) \int d\mathbf{r}' \; P_S(\mathbf{r}|\mathbf{r}', \Delta) \; \exp\left[i\gamma\delta\mathbf{g} \cdot (\mathbf{r}' - \mathbf{r})\right] , \tag{8.A.7}$$

where $P_S(\mathbf{r}|\mathbf{r}', \Delta)$ is the probability for a particle initially at $\mathbf{r}$ to end up at $\mathbf{r}'$ after time Δ, and $\rho(\mathbf{r})$ is the probability of finding the particle at $\mathbf{r}$. Assuming a spatially homogeneous $P_S(\mathbf{r}|\mathbf{r}', \Delta)$, this equation can be written in terms of the dynamic displacement $\mathbf{R} := \mathbf{r}' - \mathbf{r}$,

$$\frac{S(\mathbf{g})}{S_0} = \int d\mathbf{R} \; P_S(\mathbf{R}, \Delta) \; \exp(i\gamma\delta\mathbf{g} \cdot \mathbf{R}) \, . \tag{8.A.8}$$

Therefore, the propagator $P_S(\mathbf{R}, \Delta)$ is just the Fourier transform of the signal attenuation. With the definition $\mathbf{q} := \gamma\delta\mathbf{g}/(2\pi)$, utilization of Eq. 8.A.8 to calculate $P_S(\mathbf{R}, \Delta)$ is called q-space imaging.

Since $P_S(\mathbf{R}, \Delta)$ is physically the same function as $W_N(\mathbf{R})$ with the appropriate change of variables, so should their Fourier transforms. Therefore, the signal attenuation observed in a diffusion tensor experiment when $\delta \ll \Delta$ is equal to A_N given by Eq. 8.A.3. Therefore,

$$-\frac{1}{2}\boldsymbol{\lambda}^T \mathbf{S} \, \boldsymbol{\lambda} = -\gamma^2 g^2 \delta^2 \Delta \, \hat{\mathbf{g}}^T \mathbf{D} \, \hat{\mathbf{g}} \, . \tag{8.A.9}$$

Using this along with $\boldsymbol{\lambda} := \gamma\delta\mathbf{g}$ yields the simple relation [Özarslan and Mareci, 2003c]

$$\mathbf{D} = \frac{1}{2\Delta}\mathbf{S} \, . \tag{8.A.10}$$

Therefore, the calculated diffusion tensor, if the condition $\delta \ll \Delta$ is satisfied, is a matrix of second moments of the displacement divided by the diffusion time. And, DT-MRI can be thought of as a method that attempts to calculate this matrix without necessarily fullfilling the assumptions of q-space imaging.

References

[Bahn, 1999] Bahn, M. M. (1999). Invariant and orthonormal scalar measures derived from magnetic resonance diffusion tensor imaging. *J Magn Reson*, 141:68–77.

[Basser, 1995] Basser, P. J. (1995). Inferring microstructural features and the physiological state of tissues from diffusion-weighted images. *NMR Biomed*, 8(7-8):333–344.

[Basser, 2002] Basser, P. J. (2002). Relationships between diffusion tensor and q-space MRI. *Magn. Reson. Med.*, 47:392–397.

[Basser et al., 1994] Basser, P. J., Mattiello, J., and LeBihan, D. (1994). Estimation of the effective self-diffusion tensor from the NMR spin echo. *J. Magn. Reson.*, B(103):247–254.

[Basser et al., 2000] Basser, P. J., Pajevic, S., Pierpaoli, C., Duda, J., and Aldroubi, A. (2000). In vivo fiber tractography using DT-MRI data. *Magn. Reson. Med.*, 44:625–632.

[Beaulieu and Allen, 1994] Beaulieu, C. and Allen, P. S. (1994). Determinants of anisotropic water diffusion in nerves. *Magn Reson Med*, 31(4):394–400.

[Bloch, 1946] Bloch, F. (1946). Nuclear induction. *Phys. Rev.*, 70(7,8):460–474.

[Callaghan, 1991] Callaghan, P. T. (1991). *Principles of Nuclear Magnetic Resonance Microscopy*. Oxford: Clarendon Press.

[Chandrasekhar, 1943] Chandrasekhar, S. (1943). Stochastic problems in physics and astronomy. *Rev. Mod. Phys.*, 15(1):1–89.

[Conturo et al., 1999] Conturo, T. E., Lori, N. F., Cull, T. S., Akbudak, E., Snyder, A. Z., Shimony, J. S., McKinstry, R. C., Burton, H., and Raichle, M. E. (1999). Tracking neuronal fiber pathways in the living human brain. *Proc. Natl. Acad. Sci.*, 96:10422–10427.

[Fick and Sauermann, 1990] Fick, E. and Sauermann, G. (1990). *The Quantum Statistics of Dynamic Processes*. New York : Springer-Verlag.

[Frank, 2002] Frank, L. R. (2002). Characterization of anisotropy in high angular resolution diffusion-weighted MRI. *Magn Reson Med*, 47(6):1083–1099.

[Hahn, 1950] Hahn, E. L. (1950). Spin echoes. *Phys. Rev.*, 80:580–594.

[Hsu and Mori, 1995] Hsu, E. W. and Mori, S. (1995). Analytical expressions for the NMR apparent diffusion coefficients in an anisotropic system and a simplified method for determining fiber orientation. *Magn. Reson. Med.*, 34:194–200.

[Mori et al., 1999] Mori, S., Crain, B. J., Chacko, V. P., and van Zijl, P. C. M. (1999). Three-dimensional tracking of axonal projections in the brain by magnetic resonance imaging. *Ann. Neurol.*, 45:265–269.

[Moseley et al., 1990] Moseley, M. E., Cohen, Y., Mintorovitch, J., Chileuitt, L., Shimizu, H., Kucharczyk, J., Wendland, M. F., and Weinstein, P. R. (1990). Early detection of regional cerebral ischemia in cats: comparison of diffusion and T_2-weighted MRI and spectroscopy. *Magn. Reson. Med.*, 14:330–346.

[Özarslan et al., 2001] Özarslan, E., DeFord, S. M., Mareci, T. H., and Hayes, R. L. (2001). Mapping white matter fiber tracts with magnetic resonance imaging in rat traumatic brain injury. *J. Neurotrauma*, 18:1126.

[Özarslan and Mareci, 2002] Özarslan, E. and Mareci, T. H. (2002). Directional coherence tensor formalism for diffusion tensor MRI: Mapping structure in spinal cord gray-matter. In *Proc. of the 10th Annual Meeting of ISMRM.*

[Özarslan and Mareci, 2003a] Özarslan, E. and Mareci, T. H. (2003a). Anisotropy as a certainty measure in terms of entropy. In *Proc. of the 11th Annual Meeting of ISMRM, accepted.*

[Özarslan and Mareci, 2003b] Özarslan, E. and Mareci, T. H. (2003b). Generalized diffusion tensor imaging and analytical relationships between diffusion tensor imaging and high angular resolution diffusion imaging. *Magn Reson Med*, accepted.

[Özarslan and Mareci, 2003c] Özarslan, E. and Mareci, T. H. (2003c). On the relationships between diffusion tensor, high angular resolution diffusion and q-space imaging methods. In *44th Experimental Nuclear Magnetic Resonance Conference*, pages 267–268.

[Stejskal, 1965] Stejskal, E. O. (1965). Use of spin echoes in a pulsed magnetic-field gradient to study anisotropic, restricted diffusion and flow. *J. Chem. Phys.*, 43(10):3597–3603.

[Stejskal and Tanner, 1965] Stejskal, E. O. and Tanner, J. E. (1965). Spin diffusion measurements: Spin echoes in the presence of a time-dependent field gradient. *J. Chem. Phys.*, 42(1):288–292.

[Torrey, 1956] Torrey, H. C. (1956). Bloch equations with diffusion terms. *Phys. Rev.*, 104(3):563–565.

[Tuch et al., 1999] Tuch, D. S., Weisskoff, R. M., Belliveau, J. W., and Wedeen, V. J. (1999). High angular resolution diffusion imaging of the human brain. In *Proc. of the 7th Annual Meeting of ISMRM, Philadelphia*, page 321.

[Waxman et al., 1995] Waxman, S. G., Kocsis, J. D., and Stys, P. K. (1995). *The Axon : structure, function, and pathophysiology*. New York : Oxford University Press.

[Wedeen et al., 2000] Wedeen, V. J., Reese, T. G., Tuch, D. S., Weigel, M. R., Dou, J. G., Weiskoff, R. M., and Chessler, D. (2000). Mapping fiber orientation spectra in cerebral white matter with Fourier transform diffusion MRI. In *Proc. of the 8th Annual Meeting of ISMRM, Denver*, page 82.

[Wehrl, 1978] Wehrl, A. (1978). General properties of entropy. *Rev. Mod. Phys.*, 50(2):221–260.

Chapter 9

EPILEPTIC SEIZURE DETECTION USING DYNAMICAL PREPROCESSING (STLmax)

G. Ramachandran
Computational NeuroEngineering Laboratory, Electrical and Computer Engineering
University of Florida
Gainesville, Florida, USA, 32611
rg@cnel.ufl.edu

J.C. Principe
Computational NeuroEngineering Laboratory, Electrical and Computer Engineering
University of Florida
Gainesville, Florida, USA, 32611
principe@cnel.ufl.edu

J.C. Sackellares
Brain Dynamics Laboratory, Neuroscience
University of Florida
Gainesville, Florida, USA, 32611
sackellares@epilepsy.health.ufl.edu

Abstract The long-term properties of the Short-term largest Lyapunov exponents (STL_{max}) of EEG have been used successfully in epilepsy seizure prediction. STL_{max} profiles also show short-term patterns characterizing seizures that can be used for detection purposes. In this paper, we explore two such properties shown by the STL_{max} data during seizures and develop and compare automatic seizure detection algorithms on over 1,000 hours of data.

Keywords: STL_{max}, seizure detection, epilepsy.

1. Introduction

The design of automated seizure detectors has been addressed in the literature with many different methodologies [2, 3] and with varying degrees of success. An automatic seizure detector is useful in situations like long-term patient monitoring, clinical characterization of epilepsy, and even automated drug delivery. It is desirable to have an algorithm that is online, completely automated and patient independent.

The tasks of timing the onset, seizure identification and classification of seizures are highly subjective to electroencephalographers. This variability attests to the difficulty of the task. It is therefore no surprise that automated seizure detection algorithms are plagued by false alarms and missed detections. All of the previous attempts have utilized the EEG signal directly [2, 3]. In this paper we will investigate a dynamical pre-processor based on the short term Largest Lyapunov exponents (STLmax), which has been shown useful to predict epileptic seizures [7]. The largest Lyapunov exponent is computed in windows sufficiently long to obtain a stable estimate, but short enough to track the dynamical changes underlying neuronal populations, just like windowed FFTs can track spectral changes in time series (hence the name STLmax). Estimation of STLmax on every successive non-overlapping EEG window yields a new time series that can be used to build automated seizure detection algorithms. The hope with this dynamical preprocessor is that the many artifacts plaguing the EEG will be easier to discriminate, preserving the detectability of the seizure events and yielding better detectors. A shortcoming that we have to live with is the degradation in temporal resolution produced by the STLmax estimation. The paper is organized as follows: We start by briefly reviewing the STLmax algorithm that builds the dynamical pre-processor, and then illustrate STLmax features associated with several types of epileptic seizures. Following which we propose two methodologies to build detectors, and compare their performance on an extensive data set comprised of over 1000 hours of at least 26 channels of electrocorticographic (ECoG) data of 6 patients with temporal lobe epilepsy.

2. Nonlinear Dynamic Preprocessing

2.1. STLmax: Brief Introduction

In a series of papers [4, 5, 7], the Brain Dynamics Group at the University of Florida has proposed to model the transition from normal to epileptic state as a dynamical transition of the brain from a chaotic to a more ordered state [5, 7]. Since we do not possess a formal description of brain dynamics, dynamical parameters must be estimated directly from data. The Largest Lyapunov exponent (LLexp) appears as a good compromise between a sensitive description of relevant dynamical properties (as the transition from a lesser to a more ordered state) and effective estimation from the EEG. But since the brain, modeled as a dynamical system possesses time varying parameters, the conventional method of estimating LLexp [8] must be adapted to epileptic EEG analysis. The short-term largest Lyapunov Exponent termed STLmax has been proposed by Iasemidis and Sackellares as a means to quantify the evolving brain dynamics [4]. We briefly review the algorithm in the sequel. If we denote by L^k the estimate of the short-term largest Lyapunov exponent STLmax in k^{th} window [4, 8] then

$$L^k = \frac{1}{N_a \Delta t^k} \sum_{i=1}^{N_a} \log_2 \frac{|\delta X_{i,j}^k(\Delta t^k)|}{|\delta X_{i,j}^k(0)|} \tag{1}$$

where

- $t_i^k = t_0^k + (i-1)\Delta t^k$, $t_j^k = t_0^k + (j-1)\Delta t^k$, with $i \in [1, N_a]$ and $j \in [1, N_a]$
- Δt^k is the evolution time for $\delta X_{i,j}^k$. If the evolution time Δt^k is given in seconds, then L^k is in bit/sec
- t_0^k is the initial time point of the fiducial trajectory
- $X(t_i^k)$ is a vector of the fiducial trajectory $\phi t(X(t_0^k))$, where $t_i^k = t_0^k + (i-1)\Delta t^k$, $X(t_0^k) = [x(t_0^k), \ldots, x(t_0^k + (p-1)\tau)]^T$, and $X(t_j^k)$ is a properly chosen vector $X(t_i^k)$ in the phase space
- $\delta X_{i,j}^k(0) = X(t_i^k) - X(t_j^k)$ is a perturbation of the fiducial orbit at t_i^k and $\delta X_{i,j}^k(\Delta t^k) = X(t_i^k + \Delta t^k) - X(t_j^k + \Delta t^k)$ is the evolution of $\delta X_{i,j}^k(0)$ after time Δt^k
- N_a is the necessary number of iterations for the convergence of the L^k estimate for the data segment of N points (absolute time duration T)
- If Δt_k is the sampling period of the time domain data, then $T = (N - 1)\Delta t^k = N_a \Delta t^k - (p-1)\tau$.

The phase space reconstruction is done with an embedding dimension p of 7 and delay τ of 7 samples. The values of Δt^k and T are fixed at 20 points & 10.24 seconds respectively [4]. STLmax values are calculated for each channel independently with a non-overlapping window of 2048 ECoG samples, which makes the time scale of each STLmax point equal to 10.24 seconds.

2.2. Data Set

The development and validation of the epileptic seizure detector are conducted on a set of 6 patients with temporal lobe epilepsy (TLE). All the patients are immune to drugs and are candidates for brain surgery. The ECoG (Electro-CorticoGram) is recorded from depth electrodes as shown in Figure 9.1. Each circle represents a single electrode contact. The location of depth electrodes and subdural electrodes is depicted schematically. A_R and A_L represent subdural electrode strips placed under the right and left orbitofrontal cortex. B_R and B_L are subdural strips that are placed under the inferior temporal cortex. C_R and C_L are multi-contact depth electrodes inserted stereotactically in the right and left hippocampi respectively.

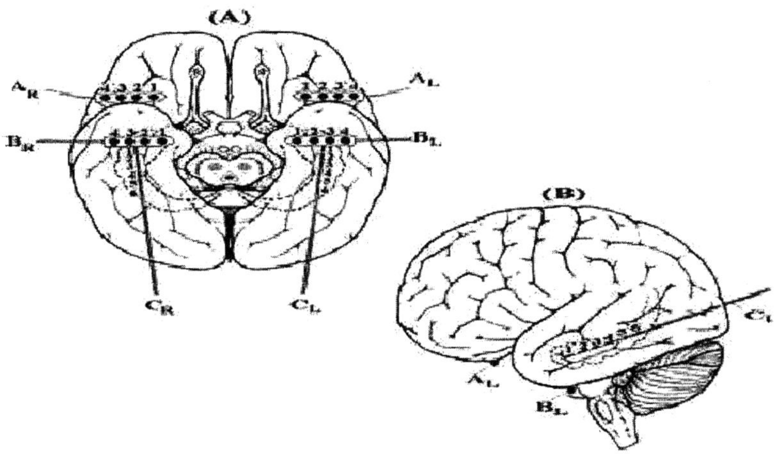

Figure 9.1. Placement of electrodes used to record the ECoG in the 6 patients considered in this work

The ECoG data from the electrodes is first digitized at 200 Hz with 10 bits precision. It is then band pass filtered at 0.1–70 Hz and recorded into VHS tapes synchronized with video recordings. Over 1000 hours of data are analyzed in this work from 6 patients with different types of seizures as summarized in Table 9.1.

Even in the case of implanted electrodes, data corruption can occur. For example, the wire connecting the electrode to the recording device can be loose or broken. The corrupted portions of the data are removed instead of being replaced with zeros, because zeros misrepresent the data. Of course concatenation also misrepresents the data as the adjoining segments may be in totally different states, though it makes the data seem continuous. We choose concatenation as it gives better results for the methods we use. Except for patient 4, the data set is continuous for all the patients.

2.3. Properties of STLmax

The data set contains four different types of seizures as scored by 2 independent neurologists [7]:

1 PSG: Complex Partial evolving to Secondarily Generalized (Figure 9.3)

2 CP: Complex Partial (Figure 9.4)

3 SC: Sub Clinical (Figure 9.5)

4 SP: Simple Partial (Figure 9.6)

The STLmax profiles show 2 patterns pertaining to all these types of epileptic seizures. They are as follows:

1 A drop in the STLmax mean value with respect to the preictal state, corroborating the hypothesis that the brain dynamics become more ordered [4]. It is interesting that this feature (a drop in value) is very different from the observed amplitude of the EEG, which normally increases during seizures. This drop can be used to detect seizures by a simple threshold mechanism.

2 Entrainment between the STLmax of different channels before a seizure [5]. In signal processing terms, this means that the "distance" between the STLmax trajectories in phase space will decrease when an ictal event is pending.

Figure 9.2 shows a typical seizure from two focal electrodes (Right Temporal Depth) where these two characteristics are clearly visible. However, it should

Table 9.1. Patient details.

Patient #	*Sex*	*Age*	*No. of Channels*	*Length of STLmax Data*		*No. of Seizures Type*				
				Hours	*No. of Breaks*	*CP*	*PSG*	*SC*	*SP*	*Total*
1	F	41	30	~211	2	19	3	1	0	23
2	M	29	28	~146	0	8	0	10	1	19
3	F	38	32	~22	0	8	0	0	0	8
4	M	60	28	~287	154	0	7	0	0	7
5	F	45	26	~87	4	3	0	6	0	9
6	M	19	28	~321	6	2	8	7	0	17
	Total			~1075	166+5	40	18	24	1	83

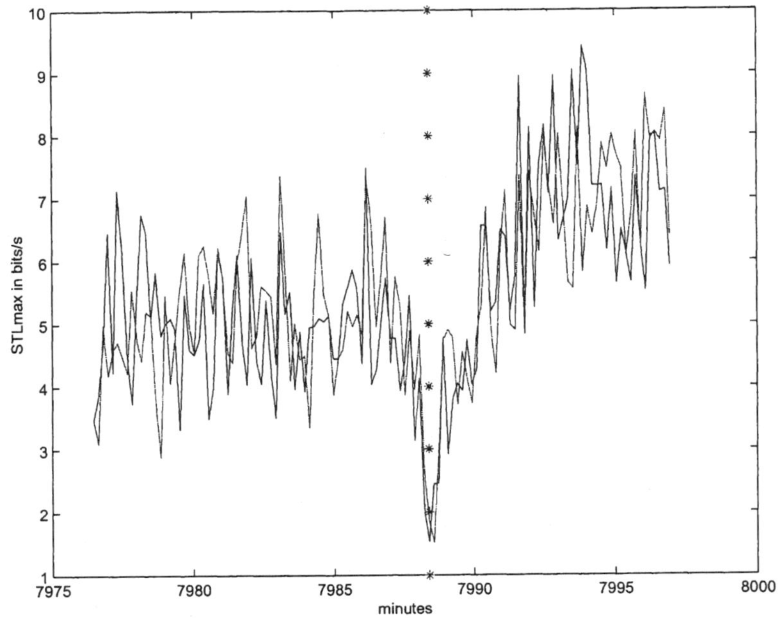

Figure 9.2. STLmax values during a seizure. The column of stars marks the seizure occurrence

be noted that the STLmax does not always show such obvious characteristics for all seizures and all electrodes, either because of inter-seizure dynamical differences, or because of seizure specific parameters to estimate STLmax. The set of parameters to estimate STLmax was held fixed throughout our experiments and works very well for complex partial seizures evolving into secondarily generalized seizures. Figure 9.3 to Figure 9.6 show examples of STLmax profiles for different seizure types.

Figure 9.3 to Figure 9.6, reinforce the observation that there are commonalities (drop in STLmax) but also differences among each type of seizure. Hence, we can expect that the level of performance in detecting a seizure will vary depending on the seizure type. Another pertinent observation is the loss of temporal resolution in the STLmax profiles. As can be observed in these figures, the seizure onset and seizure ending can only be determined with 10.24 seconds precision, and the dynamic transition itself takes several samples, and it varies from seizure to seizure. The same observation can be made for the transition to postictal state. These features imply that STLmax based detectors will provide delayed warnings in real-time applications. However, they may be more reliable than detectors applied to the ECoG directly.

The data from patient 1 is used as the training set because of the diversity of seizure types, and data from patients 2-6 are used as the test set. Table 9.2 describes the data and seizure statistics for the training and test sets.

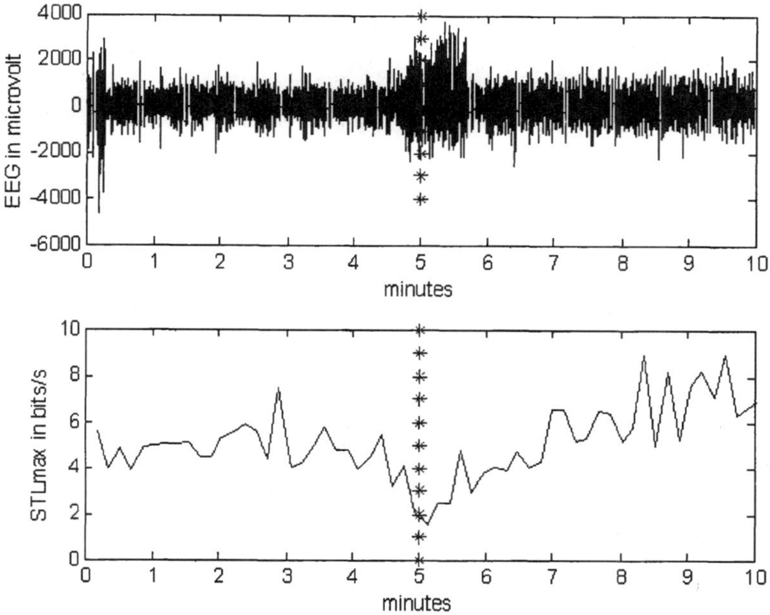

Figure 9.3. Example EEG and STLmax signatures for partial secondarily generalized seizure

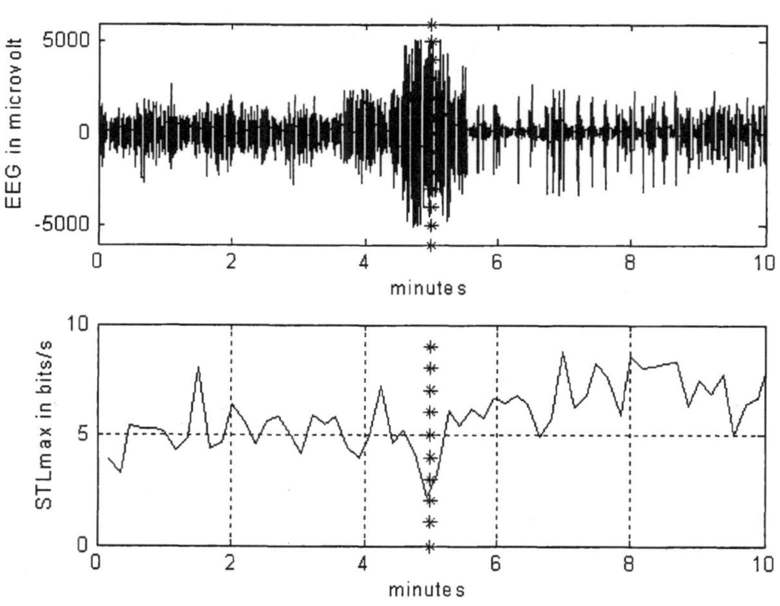

Figure 9.4. Example EEG and STLmax signatures for complex partial seizure

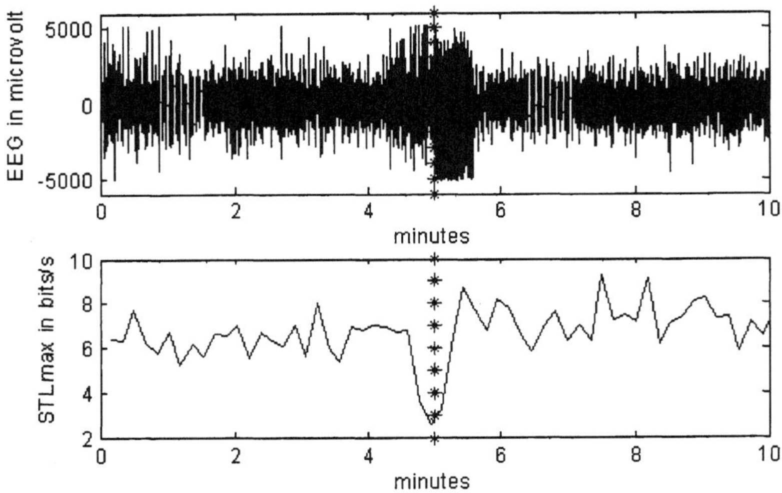

Figure 9.5. Example EEG and STLmax signatures for subclinical seizure

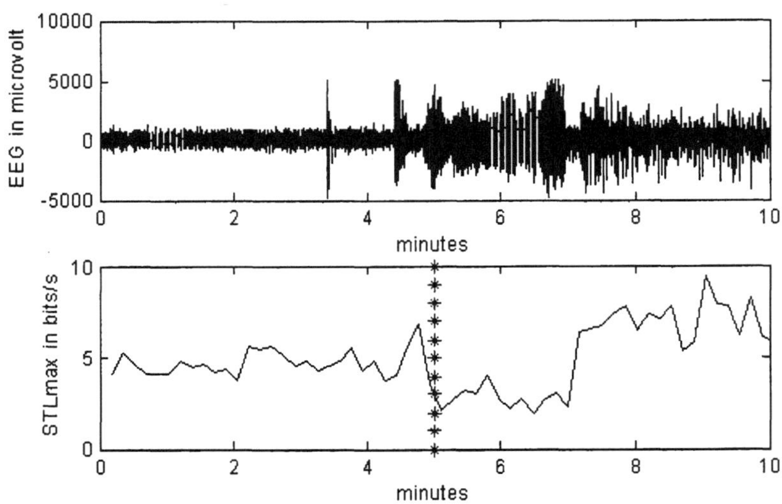

Figure 9.6. Example EEG and STLmax signatures for simple partial seizure

Table 9.2. Distribution of data among training and testing sets.

	Training set		*Test set*	
	No. of samples	*% of data set*	*No. of samples*	*% of data set*
Data length	74471	19.7	303365	80.29
No. of seizures	23	27.7	60	72.30

3. Detection Methods

We will be comparing two basic detection methods that exploit the two characteristics of STLmax profiles during a seizure. Due to the space limitations we will be presenting briefly the methods, but will avoid discussions on parameter settings, which can be found in [6].

3.1. Threshold on STLmax profiles

Observation of the STLmax profiles show that an epileptic seizure is related to a drop in STLmax. Hence, a simple detection scheme thresholds the STLmax and declares an event when the signal goes below the threshold. Since we have multiple channels of STLmax data, there are a number of possible combinations in which the multi-channel STLmax time series can be used. For example, the detection mechanism can operate on a subset of the channels, as follows:

1 a drop in the minimum STLmax among the channels

2 a drop in the average STLmax among the channels

3 a drop in the maximum STLmax among the channels

4 a drop in the median STLmax among a subset of channels

5 a drop in the variance among the channels

Alternatively, the detection can be done in each of the channels independently with a voting scheme, e.g. a drop in a critical mass of channels. Our hypothesis is that, for a seizure to happen, there might be a critical mass of brain areas that display a drop in the STLmax.

3.2. Entrainment in STLmax and Seizure Detection

The second characteristic observed before a seizure, is the co-variability of the STLmax profiles among channels and convergence to a small interval of values, which we call the entrainment [4]. One possibility to measure entrainment would be the correlation coefficient, but this is a linear measure and it is limited to channel pairs. We decided to implement the Diks distance proposed recently to measure the difference between attractors in chaotic systems [1]. Using Diks distance one can create a statistic that yields the Diks test. A seizure will then be translated by a decrease in Diks distance just prior and during seizure.

3.2.1 Brief introduction: Diks test and Diks distance. In simple words, Diks test [1] measures the distance between two multidimensional probability distributions, constructed from the delay vectors, and normalized by its variance.

Consider the case of N_1 vectors $\{\overrightarrow{X_i}\}_{i=1}^{N_1}$ ($\overrightarrow{X} = [x(n), x(n-\tau), \ldots, x(n-(m-1)\tau)]$), with probability distribution $\rho_1(\overrightarrow{X_i})$, and N_2 vectors $\{\overrightarrow{Y_i}\}_{i=1}^{N_2}$, with probability distribution $\rho_2(\overrightarrow{Y_i})$. Realizations of these vectors are denoted by their lowercase analogs. The smoothed versions of the multidimensional distributions (ρ'_k) are constructed as

$$\rho'_k(\overrightarrow{r}) = \int \mathrm{d}(\overrightarrow{s})\, \rho'_k(\overrightarrow{s})\kappa(\overrightarrow{r}, \overrightarrow{s}) \qquad \text{for } k \in \{1, 2\} \tag{2}$$

where $\kappa(\overrightarrow{r}, \overrightarrow{s})$ is the Gaussian kernel defined as

$$\kappa(\overrightarrow{r}, \overrightarrow{s}) = \left(\sqrt{2\pi}d\right)^{-m} e^{-|\overrightarrow{r} - \overrightarrow{s}|/2d^2} \tag{3}$$

where $d > 0$ is the bandwidth (kernel standard deviation).

An unbiased estimator $\hat{\rho}'_k(\overrightarrow{r})$ of $\rho'_k(\overrightarrow{r})$ is

$$\hat{\rho}'_k(\overrightarrow{r}) = \frac{1}{N_1}\sum_{i=1}^{N_1} \kappa(\overrightarrow{r}, \overrightarrow{x_i}) \tag{4}$$

since the expected value of $\kappa(\overrightarrow{r}, \overrightarrow{s})$ is,

$$\int \mathrm{d}(\overrightarrow{X_i})\, \rho_1(\overrightarrow{X_i})\kappa(\overrightarrow{r}, \overrightarrow{X_i}) = \rho'_1(\overrightarrow{r}) \tag{5}$$

We know that the Euclidean distance between the distributions can be estimated as,

$$Q = \left(2d\sqrt{\pi}\right)^m \int \mathrm{d}\overrightarrow{r}\, [\rho'_1(\overrightarrow{r}) - \rho'_2(\overrightarrow{r})]^2 \tag{6}$$

More details about the properties of the measure are presented in [1]. It is claimed that the above distance becomes zero if and only if the two distributions are identical. By determining whether a consistent estimator of Q is significantly above zero we can test the null hypothesis $\rho_1 = \rho_2$ against all alternatives $\rho_1 \neq \rho_2$.

From the training set we selected an embedding dimension of 6, and a delay of 7 to reconstruct the attractors, and used 60 sample segments with 59 sample overlap from the STLmax time series. The Gaussian kernel bandwidth was chosen as one. Figure 9.7 displays the Diks distance for a seizure, where it is well apparent that in fact the distances between all the channel pairs decrease towards and during this seizure. The delay between the trough and the time of seizure occurrence is corresponds to the number of samples needed to estimate the distance with reasonable accuracy.

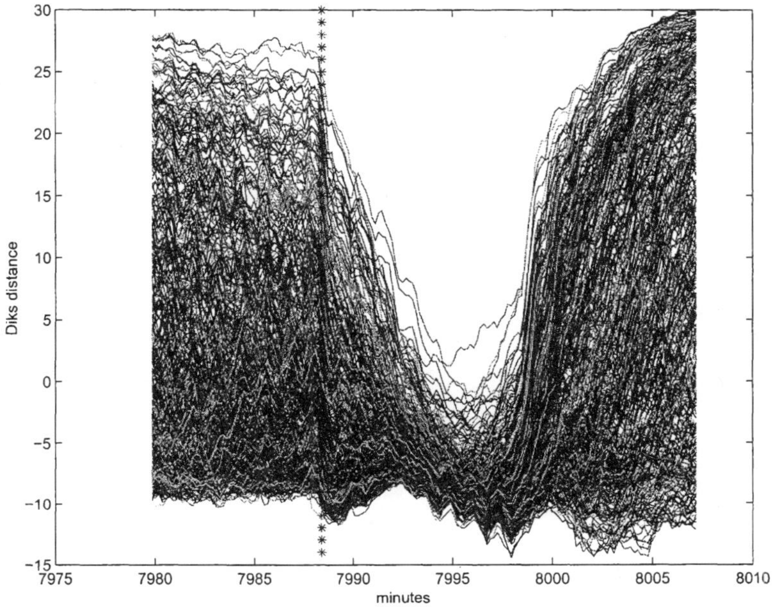

Figure 9.7. Diks distances for all the possible pair wise combination of electrodes during a seizure.

4. Results and Discussion

4.1. Definitions and Trade-offs

In this chapter, we will present results of the two detection methodologies. However, first we have to specify the details of the definitions and validation. In order to compare different methods, the following clinical relevant receiver operating characteristics (ROC) criteria are used:

1 Percentage of seizures detected. Along with it, we also enforce maximum amount of delay between the seizure onset and the instant the algorithm declares it as a seizure (Figure 9.8).

2 Number of false alarms per hour. We restrict for clinical acceptability the number of false alarms at 1 per hour. However, to avoid multiple triggering in the same event we will enforce a "dead period" after detection.

The following definitions are used to determine the above criteria:

Hit – The alarm turns ON within the maximum allowed delay from the seizure onset.

We still need to define when the alarm turns off. From Figure 9.9, we can see that this particular threshold will produce two alarms for the same seizure.

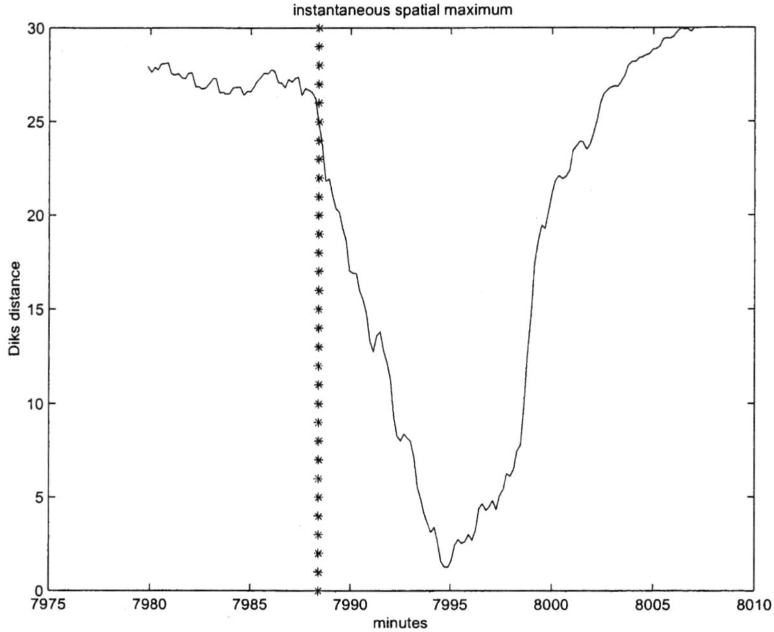

Figure 9.8. Figure showing the delay between the instant of the actual event to the instant of the alarm.

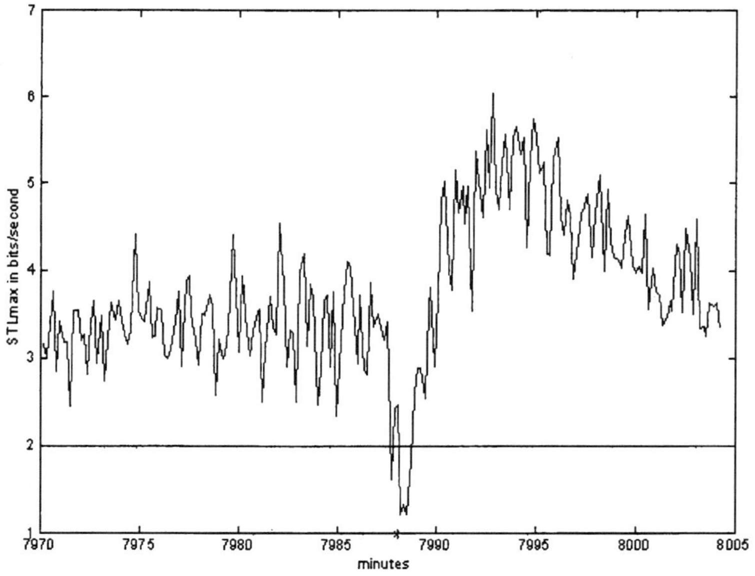

Figure 9.9. Subset of STLmax showing the same seizure producing two alarms.

Using our definitions above, we would count a true detection and a false alarm for this event, which is not right.

We can avoid such situation by clustering the alarms. In the data set under study, the shortest interval between successive seizures is 18 minutes. Therefore, all alarms that occur within a span of 18 minutes can be clustered into a single detection (with the timing of the first alarm) without having to fear about one alarm representing more than a seizure. Implicitly we are constraining the resolution of the algorithm i.e., the minimum amount of time the algorithm requires between seizures to identify them separately. In this work, a conservative estimate of 15 minutes is used. Also if an alarm lasts for more than 15 minutes, it is reset after 15 minutes and counted as more than one alarm. Table 9.3 describes the mnemonics used in the ROC plots. It is to be noted that the median is applied on a subset of channels. The each instant. Therefore, it is presented separately as it has one extra variable N.

4.2. Test Set Results

We first present the ROC results as a function of how fast the alarm is turned on after the occurrence of the true event (which has been determined by trained neurologists). To improve the visibility of plots, only the best results for each subset of the detectors (STLmax, Diks distance) for each false alarm rate are shown. To make the comparisons easier, the axes are set to show detection rates from 0% to 100% and the false alarm rates from 0.03/h to 1/h. From Figure 9.10, we can see that STLmax works the best for fast detectors (delay of 1 minute) for all the false alarm rates except near 1/h. However, the performance is still not acceptable with at most a detection rate of $\sim$55% for the minimum of Diks distance. When the allowed delay in detection is increased to 5 minutes, not only the overall performance of all the detectors improve, but also the best performer becomes the minimum of Diks distance with a detection rate of 82%. We expected that Diks distance would be the best performer at intermediate lags, because it measures entrainment that starts occurring before the seizure. However, it requires a window of data to be estimated properly, and at 5 minutes, it performs better than subsets of STLmax for most cases.

Table 9.3. Table of mnemonics.

Signal	*Subset*			
	Minimum	*Average*	*Maximum*	*Variance*
STLmax	minlmax	meanlmax	maxlmax	lmaxvar
Diks	mindiks	meandiks	maxdiks	diksvar

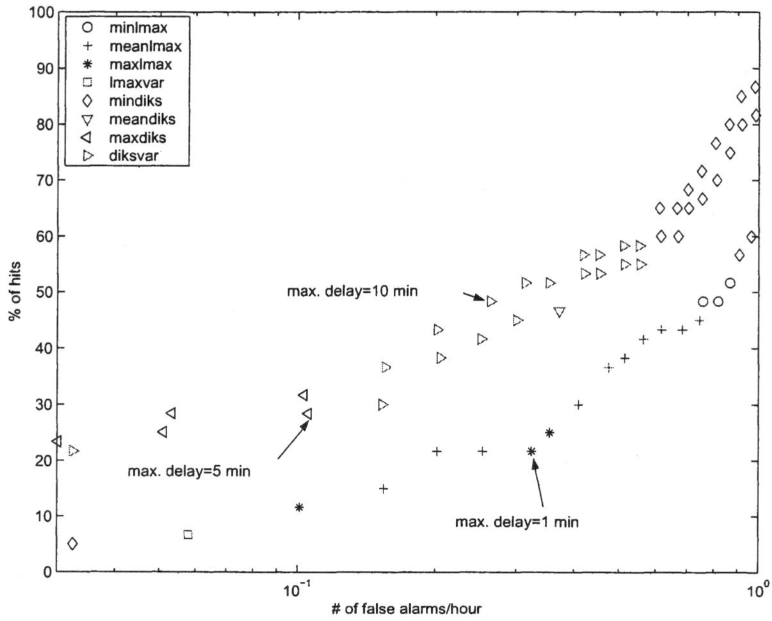

Figure 9.10. ROCs of subsets on the test set.

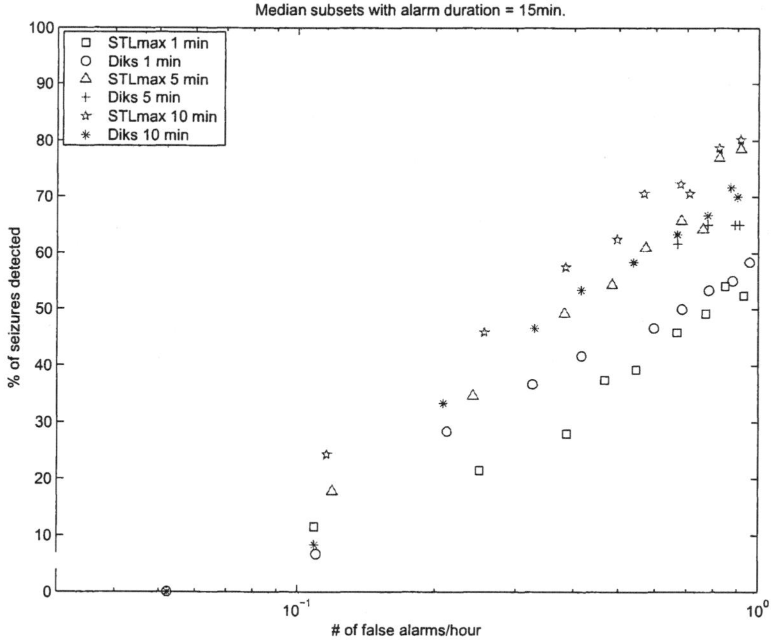

Figure 9.11. ROCs of median subsets on the test set.

For 10 minute delay between the seizure onset and the alarm, the best performer is still the minimum of Diks distances and the figure of merit is rather high, 87% sensitivity for 1 false alarm per hour. This value seems far better than the reported performance of seizure detectors based on EEG [2, 3], but it requires a direct comparison in our data set.

We next analyze the differences in performance among the multi-channel decision rules and the voting. The voting scheme performs a little poorer than their counterparts (around 2-5% less sensitivity (detection rate) for the same false alarm rate). The main reason is that, although there might be a critical mass mechanism operating during seizures, the amount of brain involved may be different from seizure to seizure. With respect to the median operator, it performed a little better in the case of STLmax, but it was poorer in the case of Diks distance. Further discussion will be omitted here.

4.3. Case studies: Best Performance on the Test Set

4.3.1 Quick Detector: STLmax. From Figure 9.10 to Figure 9.12, for short delays, we can see that the spatial median of STLmax and spatial minimum of Diks distance perform better than the others. Since the former has lesser computational complexity than the latter, it is chosen to be studied in terms of the seizure types, keeping the maximum delay allowed at 1 minute as well as the maximum false alarm rate at 1 per hour.

Figure 9.13 presents the performance of the median operator for different values of N, the number of channels. We can see that a subset of 13 channels gives best performance; therefore, it will be analyzed in detail. From Figure 9.14, we can see that clinical seizures are detected better than sub-clinical ones. Also, we can see that the algorithm performs best for secondarily generalized complex partial seizures.

4.3.2 Slow Detector: Spatial minimum of Diks distance. From Figure 9.10 and Figure 9.12, for long delays, we can see that spatial minimum of Diks distance produces very good results (almost 90%). Analyzing it in detail, from Figure 9.15, we can see that it performs well for clinical seizures (both generalized and local). One of the reasons for this phenomenon may be due to the fact that the training set consists mostly of clinical seizures.

5. Conclusion and Directions for Future Work

In this work, two different detection methods are studied each focusing on one of the characteristics of the STLmax signature of a seizure. The results are summarized in Table 9.4.

For quick response detectors, the STLmax thresholding performed better than Diks distance. This is expected due to the inherent delay in the computa-

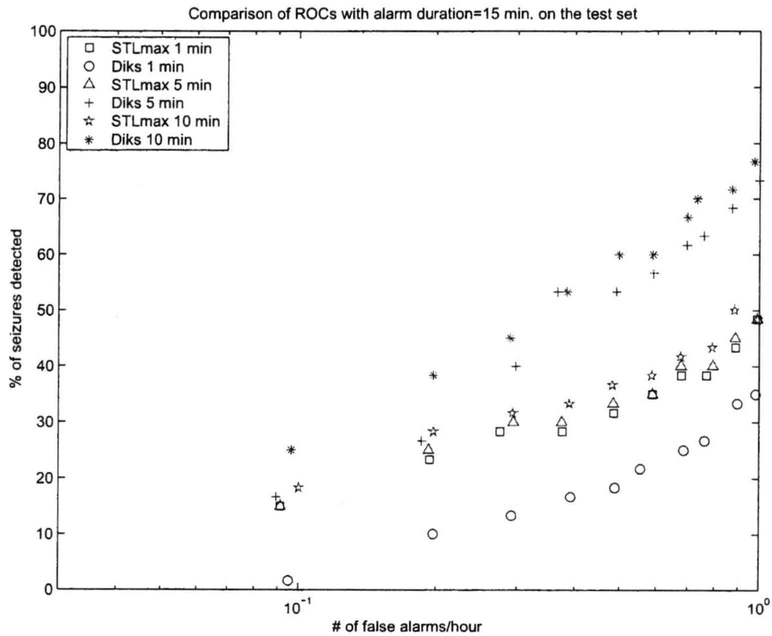

Figure 9.12. ROCs of voting on the test set.

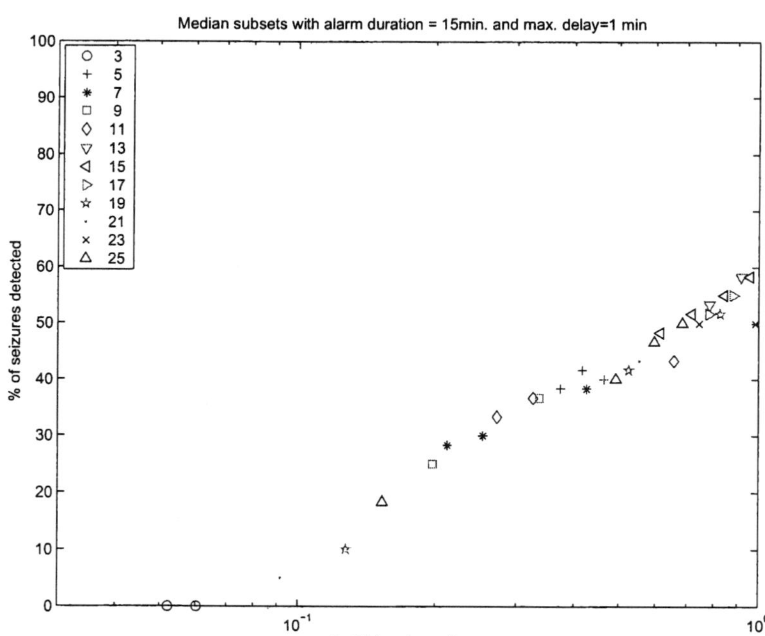

Figure 9.13. ROC of median subsets for a maximum allowable delay of 1 minute on the test set.

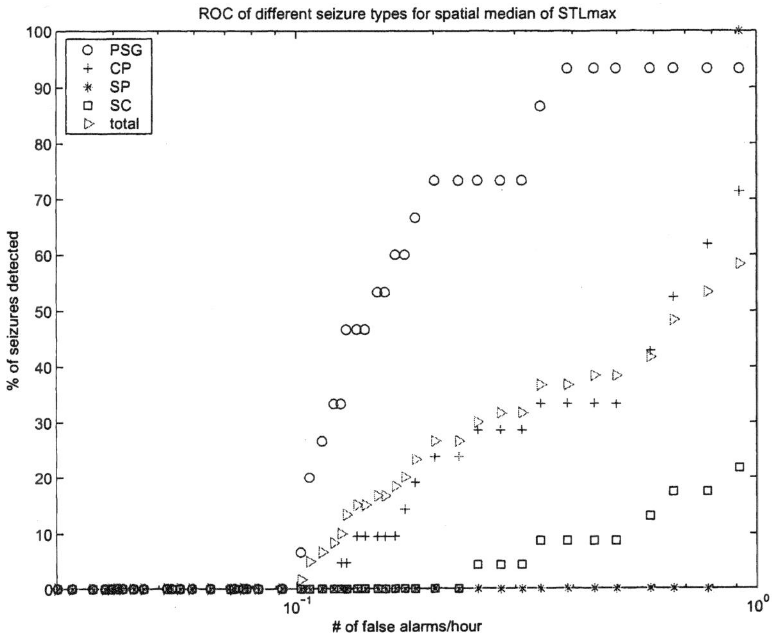

Figure 9.14. ROC of different seizure types using spatial median of STLmax for a maximum allowable delay of 1 minute on the test set.

Table 9.4. Best detection rates on test set for false alarm rate $< 1/h$ (ordered by result).

	Delay	
1	5	10
Median STLmax (58.33)	Min. Diks (81.67)	Min. Diks (86.67)
Min. Diks (58.33)	Median Diks (78.52)	Median Diks (80.19)
Median Diks (54.07)	STLmax vote (76.67)	STLmax vote (78.33)
Min. STLmax (53.33)	Median STLmax (65.00)	Median STLmax (71.67)
Mean STLmax (53.33)	Min. STLmax (63.33)	Var. Diks (70.00)
Diks vote (50.00)	Mean STLmax (60.00)	Mean Diks (65.00)
Max. STLmax(41.67)	Max. Diks (60.00)	Min. STLmax (65.00)
STLmax vote (36.67)	Var. Diks (58.33)	Max. Diks (61.67)
Var. STLmax (36.67)	Mean Diks (56.67)	Mean STLmax (60.00)
Mean Diks (28.33)	Diks vote (50.00)	Diks vote (53.33)
Max. Diks (28.33)	Max. STLmax (43.33)	Max. STLmax (43.33)
Var. Diks (26.67)	Var. STLmax (36.67)	Var. STLmax (36.67)

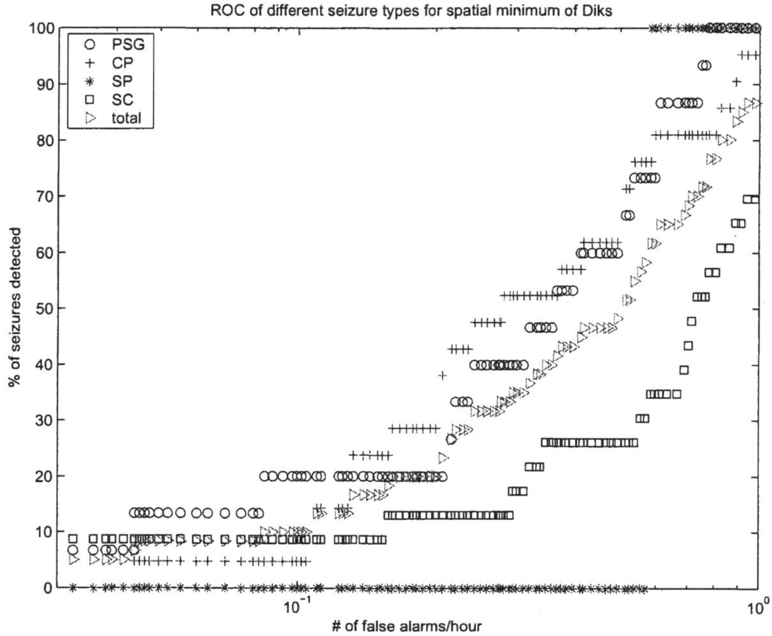

Figure 9.15. ROC of minimum Diks distance for a maximum allowable delay of 1 minute on the test set.

tion of Diks distance. Also, subsets of STLmax performed better than others in the case of secondarily generalized seizures. The minimum of Diks distance gave a good generalization for all types of seizures under study. Due to the simplicity of the computation, unprocessed STLmax is ideally suited for real time implementation. Overall, we think that the performance in terms of detectability (87%) for one false alarm per hour makes STLmax a very interesting preprocessor for automated seizure detection in long term monitoring of epileptic patients. Obviously, the downside is the delay between seizure occurrence and alarms, which must be at least 5 minutes. At this point it precludes the use of these detectors for real time applications such as drug intervention or warning. However, if the problem is one of transient response, we can overlap the windows for STLmax estimation and decrease the amount of time the Diks distance need to converge to their final values among preictal and ictal events. More research needs to be conducted in this aspect.

All the parameters were found using the training set, and of course, they may not be the optimal ones for the test set. Also, the use of Diks distance is based on the assumption of entrainment between electrodes, which may not be true for all types of seizures. This reasoning means that we may have to adapt these parameters for each patient to improve performance. Another aspect that requires further analysis is the best combination of all these detectors. Due to the sheer amount of parameters, a genetic search algorithm may be required.

Acknowledgments

The authors will like to thank Dr. Paul Carney and Dr. Deng S. Shiau for their assistance in understanding the epilepsy data and the medical concepts involved.

References

[1] C. Diks, W. R. Van Zwet, F. Takens, and J. De Goede. Detecting differences between delay vector distributions. *Physical Review E*, 53(2):2169–2176, 1996.

[2] J. Gotman and P. Gloor. Automatic recognition and quantification of interictal epileptic activity in the human scalp EEG. *Electroencephalography and Neurophysiology*, 41:513–529, 1976.

[3] J. Gotman and P. Gloor. Automated seizure detection using a self-organizing neural network. *Electroencephalography and Clinical Neurophysiology*, 99:257–266, 1996.

[4] L. D. Iasemidis and J. C. Sackellares. The evolution with time of the spatial distribution of the largest lyapunov exponent on the human epileptic cortex. In D. Duke and W. Pritchard, editors, *Measuring Chaos in the Human Brain*. World Scientific, Singapore, 1991.

[5] L. D. Iasemidis and J. C. Sackellares. Chaos theory and epilepsy. *Neuroscientist*, 2:118–126, 1996.

[6] G. Ramachandran. Comparison of algorithms for epileptic seizure detection. Masters Thesis, University of Florida, Gainesville, FL, USA.

[7] J.C. Sackellares, L.D. Iasemidis, D. S. Shiau, R. L. Gilmore, and S. N. Roper. Epilepsy – when chaos fails. In K. Lehnertz, J. Arnhold, P. Grassberger, and C. E. Elger, editors, *Chaos in the Brain*, page 112. World Scientific, Singapore, 2000.

[8] A. Wolf, J. B. Swift, H. L. Swinney, and J. A. Vastano. Determining lyapunov exponents from a time series. *Physica D.*, 16:285–317, 1985.

Chapter 10

ROLE OF THE DORSOCENTRAL STRIATUM IN CONTRALATERAL NEGLECT AND RECOVERY FROM NEGLECT IN RATS

Roger L. Reep
Department of Physiological Sciences and McKnight Brain Institute
University of Florida
Gainesville, Florida, USA, 32610
reep@ufbi.ufl.edu

J.L. Cheatwood
Department of Physiological Sciences and McKnight Brain Institute
University of Florida
Gainesville, Florida, USA, 32610
cheatwoo@ufl.edu

J.V. Corwin
Department of Psychology
Northern Illinois University
DeKalb, Illinois, USA, 60115
jcorwin@niu.edu

Introduction

Neglect is a complex and devastating human neuropsychological disorder characterized by a failure to attend to novel or meaningful stimuli presented to the side contralateral to a brain lesion, in the absence of a primary sensory or motor dysfunction. Some manifestation of neglect is found in approximately 40% of all cases of brain damage, although more frequently following right hemisphere damage. The vast majority (80-90%) of all cases of neglect are produced by destruction of one of three cortical regions: the dorsolateral prefrontal cortex, the cingulate cortex, or the parietotemporal junction [Heilman et al., 1993,Karnath et al., 2001,Mesulam, 1980,Mesulam, 1990]. The neglect syndrome goes far beyond a lack of responsiveness to contralesional stimuli to include dramatic attentional and cognitive spatial deficits. These include several major associated disorders. Patients may exhibit hemispatial neglect in which the patient does not have a cognitive representation of the contralesional hemispace, and will omit details of an imagined scene. Hemiinattention to contralesional stimulation across several or one sensory modality is also quite common. The patients may display hemiakinesia, or a paucity of movement of the contralesional limbs. Inappropriate orientations (allesthesia) may also occur in which the patient will respond or orient to contralesional stimulation as if it came from the ipsilesional side. Extinction to bilateral simultaneous stimulation may also be exhibited. Often during the course of recovery from severe neglect, when the patient is stimulated on the contralesional side alone they may respond to the stimulation, but if stimulated bilaterally they may only demonstrate awareness of the ipsilesional stimulus. Further, these patients may exhibit anosognosia, a lack of awareness of their current neurological state, or anosodiaphoria in which they are unconcerned about their neurological status [Heilman et al., 1993].

As a consequence, patients may fail to recognize the limbs on the contralesional side as their own, eat off of one side of a plate, and groom only their ipsilesional side. The cognitive, affective, and motor disturbances that comprise the neglect syndrome are often debilitating to the patients and their families [Robertson et al., 1993]. The failure to respond to contralesional stimuli has devastating effects for the patients' abilities to demonstrate independent daily living or to return to work, and the presence of neglect in patients has been found to be the best predictor of a poor prognosis for recovery [Denes et al., 1982,Fullerton et al., 1997,Kerkhoff, 2001]. Further, neglect has been shown to be related to poor recovery from related deficits such as hemiplegia [Denes et al., 1982].

Recovery, when it occurs, is spontaneous over the course of weeks to months, but often it is incomplete. Many neglect patients, perhaps as high as 15-25% [Kerkhoff, 2001], continue to ignore or neglect contralesional stimuli for months or years post lesion [Halligan and Marshall, 1994,Henley et al., 1985,Kerkhoff, 2001]. They are often unaware of their neurological status (anosognosia), or are unconcerned and affectively flat (anosodiaphoria) [Heilman et al., 1993]. The lack of awareness and changes in motivation interfere with successful physical and occupational rehabilitation [Robertson et al., 1993]. Behavioral treatments rarely generalize outside of the therapeutic context or across tasks within the same therapeutic context [Couvier et al., 1987, Robertson et al., 1993]. Drug therapies are rarely used to treat neglect because there has been no rational framework for understanding the mechanisms which might lead to a therapeutic effect. Typically, drugs have been given only to patients with chronic neglect with stable behavioral baselines, because of the concern that drug effects may interfere with ongoing recovery [Fleet et al., 1987, Geminiani et al., 1998, Hurford et al., 1998].

1. The Rodent Model of Neglect and Recovery

The poor prognosis and the absence of generally accepted therapies for neglect have led to the development of animal models of neglect [Milner, 1987]. We have developed a rat model of neglect to examine the basic mechanisms of neglect, and the potential for recovery of function. A rodent model offers several practical advantages in terms of accessibility and cost over many of the primate models, and we have found striking behavioral, pharmacological and anatomical similarities between the systems related to neglect in rodents and those in primates [Corwin and Reep, 1998].

We [Burcham et al., 1997, Corwin and Reep, 1998] have found that in rats, as in primates [Heilman et al., 1993], there is a cortical network for directed attention, involving the medial agranular (AGm), posterior parietal (PPC), and ventrolateral orbital cortices. The symptoms of neglect in rodents are similar to those found in human neglect patients including: severe neglect of visual, tactile, and auditory stimuli presented in the hemispace contralateral to the lesion, extinction, allesthesia, and disorders of spatial processing [King and Corwin, 1992, King and Corwin, 1993, Corwin and Reep, 1998]. Severe neglect is defined as the contralesional side being only 33% as responsive as the ipsilesional side. The tests for neglect and extinction in rodents were designed to mirror the bedside testing of neglect patients [Crowne et al., 1983, Corwin et al., 1986, Van Vleet et al., 2000, Van Vleet et al., 2002]. Selective disconnection of the corticocortical axons linking AGm and PPC in the absence of direct damage to either area results in neglect, demonstrating that it is the integrity of the network that is critical for normal directed attention [Burcham et al., 1997].

Studies of recovery from neglect in rats have focused mainly on area AGm, and the results indicate a number of strong similarities to recovery in humans. In rats recovery from AGm-induced neglect has been found to occur in three contexts: (1) as found in humans, rodents with neglect demonstrate severe long-term deficits with some limited spontaneous recovery which may occur over the course of weeks to months [Corwin et al., 1986, King and Corwin, 1990, King and Corwin, 1993], (2) as in humans dopamine (DA) agonists produce acute recovery of function [Corwin et al., 1986, Fleet et al., 1987, Geminiani et al., 1998, Hurford et al., 1998, King and Corwin, 1990, Van Vleet et al., 2003b], and (3) exposure to 48 hr of light deprivation (LD) at 4 hr postsurgery produces dramatic immediate recovery [Burcham and Corwin, 1998, Crowne et al., 1983, Van Vleet et al., 2003a].

2. Role of the Dorsocentral Striatum in Neglect

2.1. Behavioral Studies

While it was documented that recovery does occur in the three contexts mentioned above, the crucial systems involved in recovery and the neural mechanism remained unknown. However, studies in rodents and humans indicated that the striatum may play a crucial role in neglect and recovery from neglect produced by unilateral AGm lesions.

The striatum receives glutamatergic inputs from cortex and thalamus, and dopaminergic inputs from substantia nigra (SN) that converge on the same medium spiny neurons [Parent, 1990]. Neglect can be produced by cortical lesions, or disruption of the nigrostriatal DA system [Marshall and Gotthelf, 1979], or by striatal infarcts [Caplan et al., 1990]. The importance of the striatum in AGm lesion-induced neglect was originally suspected because systemic delivery of apomorphine (a DA receptor agonist) can produce acute recovery [Corwin et al., 1986, King and Corwin, 1990], and spiroperidol (a DA receptor antagonist) can reinstate neglect [Vargo et al., 1989]. Based on our findings and those of Marshall on subcortical neglect in rats, Fleet et al. [Fleet et al., 1987], Geminiani et al. [Geminiani et al., 1998], and Hurford et al. [Hurford et al., 1998] examined the effects of DA agonists (bromocriptine or apomorphine) in human patients with chronic neglect and found that DA agonists produced significant acute recovery. While DA agonist treatment is effective, the site of action for the therapeutic effects was unknown.

Vargo and Marshall [1996a; 1996b] were the first to suggest that plastic changes in the striatum were correlated with neglect and behavioral recovery from AGm-induced neglect. They found that neglect was correlated with decreases in NMDA and kainate receptors in the ipsilesional dorsolateral striatum, and recovery was correlated with a normalization of kainate receptors and a 10% increase in NMDA receptors in this region. The basis for the changes in glutamate receptors was unknown, and could be based on changes in presynaptic or postsynaptic receptors [Chen and Hillman, 1990].

Based on these findings we began a series of behavioral, pharmacological, and anatomical studies to examine the role of the striatum in neglect and behavioral recovery of function. Based on our anatomical findings [Reep and Corwin, 1999] the major site for the projections of the AGm was a region in the dorsocentral striatum (DCS) (Figure 10.1), therefore the initial studies were designed to examine the potential role of this region.

In the initial study of the DCS we directly compared the behavioral effects of unilateral lesions of the AGm or the DCS on the major manifestations of the neglect syndrome. The results of this study indicated that axon-sparing unilateral DCS lesions resulted in severe multimodal neglect (Figures 10.2 and 10.3) which did not spontaneously recover even after 96 days of testing, thus

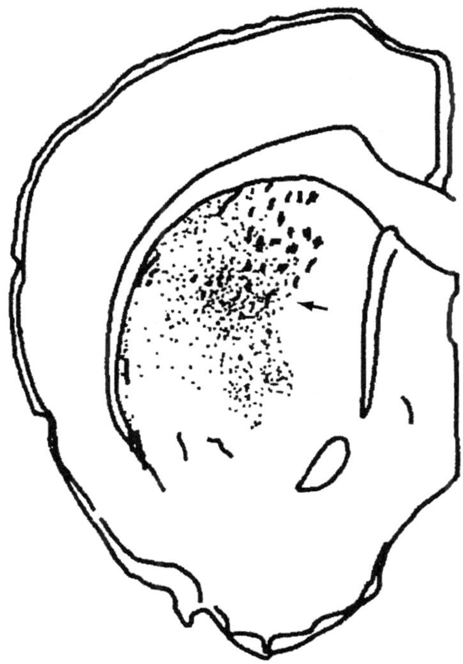

Figure 10.1. In this coronal section through the forebrain, labeled axons from cortical area AGm terminate most densely in the dorsocentral region of the striatum (arrow). There is also a substantial projection to the margin of the striatum bordering the external capsule. Thick profiles represent axon fascicles.

supporting the contention that the DCS is a critical component of the network for neglect and directed attention [Van Vleet et al., 2000]. However, unlike AGm lesions, unilateral DCS lesions did not produce extinction or allesthesia [Van Vleet et al., 2002] (Figure 10.4). The results were the first to indicate that neglect and extinction are experimentally dissociable. Further, the results indicated that extinction deficits may be based on a disruption of cortical systems related to directed attention. It is of some interest that neither AGm or DCS - induced neglect were correlated with increases in circling behavior which indicates that the deficits in orientation were not secondary to postural or motor asymmetries. These results are of some import because they suggest that the deficits associated with the neglect syndrome are multiple. However, to the degree they are dissociable it may be possible to target prospective behavioral or pharmacological treatments so as to deal with specific individual deficits [Pyter et al., 2002]. These considerations are clinically relevant because extinction deficits often persist in patients that have recovered spontaneously from neglect [Kerkhoff, 2001].

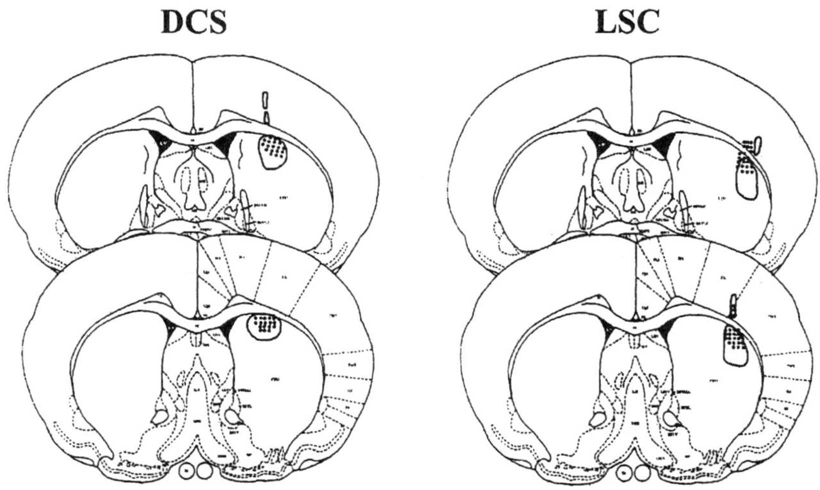

Figure 10.2. Axon-sparing lesions of the dorsocentral striatum, made using NMDA. Maximum (outlined) and minimum (stippled) extents of the lesions are shown at two a-p levels for the DCS and lateral striatal control (LSC) groups.

In a subsequent studies the role of the DCS in acute drug-induced recovery was examined. The results of these studies indicated that the likely site of action for the therapeutic effects of apomorphine on neglect is the DCS. Van Vleet et al. [Van Vleet et al., 2000] found that apomorphine was ineffective in subjects with neglect induced by unilateral destruction of the DCS. These findings led to the conclusion that the integrity of the DCS was necessary for the therapeutic effects of apomorphine. In order to examine the role of the DCS as the crucial site for the therapeutic effects of apomorphine we infused apomorphine directly into the DCS or a laterally adjacent area not implicated in the circuitry related to neglect. Apomorphine infusion into the DCS produced dramatic dose-dependent recovery from multimodal neglect [Van Vleet et al., 2003b]. Infusion into a laterally adjacent area in the striatum did not produce a therapeutic effect, even when much higher dosages were used (Figure 10.3). The drug-induced recovery was virtually identical to that found in prior studies which used systemic administration of apomorphine [Corwin et al., 1986, King and Corwin, 1990]. We have also examined the effects of apomorphine on extinction [Pyter et al., 2002]. We have found that, in keeping with previous studies in rats [Corwin et al., 1986, King and Corwin, 1990] and humans [Fleet et al., 1987, Geminiani et al., 1998, Hurford et al., 1998] dopamine agonists produce a therapeutic effect on severe multimodal neglect. However, apomorphine does not produce a therapeutic effect on extinction in these same subjects.

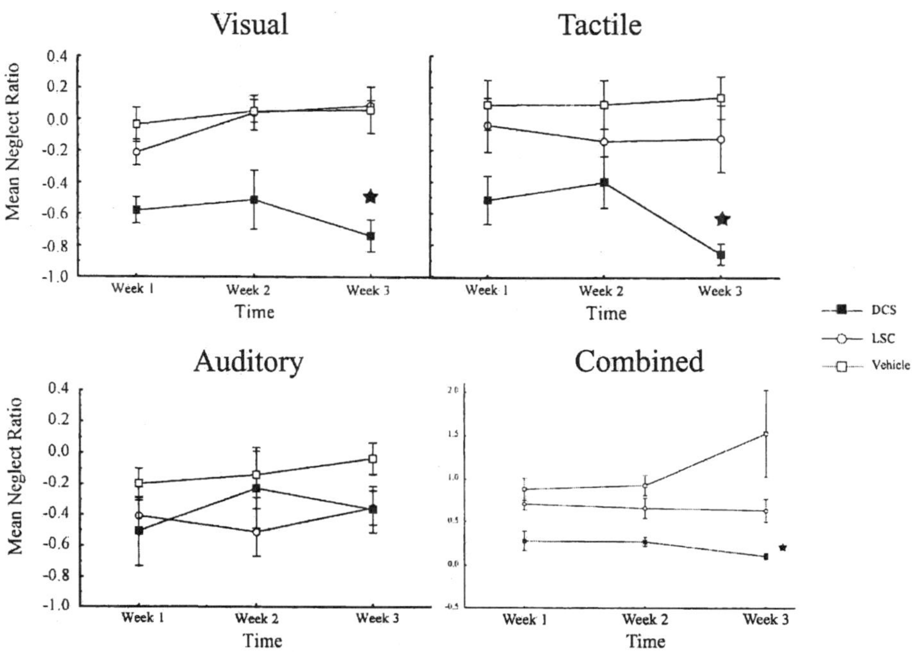

Figure 10.3. Neglect ratios for the DCS-lesioned group, vehicle control DCS group, and LSC lesion control group. A neglect ratio of 1.0 represents normal, symmetrical responding to stimuli presented on the left or right side; ratios < 0.4 represent severe neglect. Data are presented for each modality separately as well as in a total neglect ratio that combines the data from all modalities. Stars represent significant differences on a weekly basis. The DCS-lesioned group had a significantly ($P < 0.05$) lower total neglect ratio than the other two groups when data were collapsed across weeks.

Recovery from AGm-induced neglect can be produced by exposing the AGm operates to 48 hr of light deprivation within 4 hr of the lesion. The results of a recent study from our group have indicated that the DCS is the likely site of action for the therapeutic effects of light deprivation [Van Vleet et al., 2003a]. In that study, subjects with combined unilateral AGm and DCS lesions did not demonstrate light deprivation-induced recovery, while subjects with AGm lesions combined with a lesion of the striatum laterally adjacent to the DCS did demonstrate dramatic recovery. In this study we also found that the therapeutic effects of light deprivation were limited to neglect; a severe extinction deficit was unaffected by light deprivation. These results further support the contention that the neural substrates for neglect and extinction are dissociable and that the substrates which underlie recovery are likely to differ as well.

The results of the behavioral studies provide strong evidence that the DCS is an crucial component of the circuitry related to directed attention and neglect. Further, the integrity of the DCS is crucial for recovery from neglect in all three

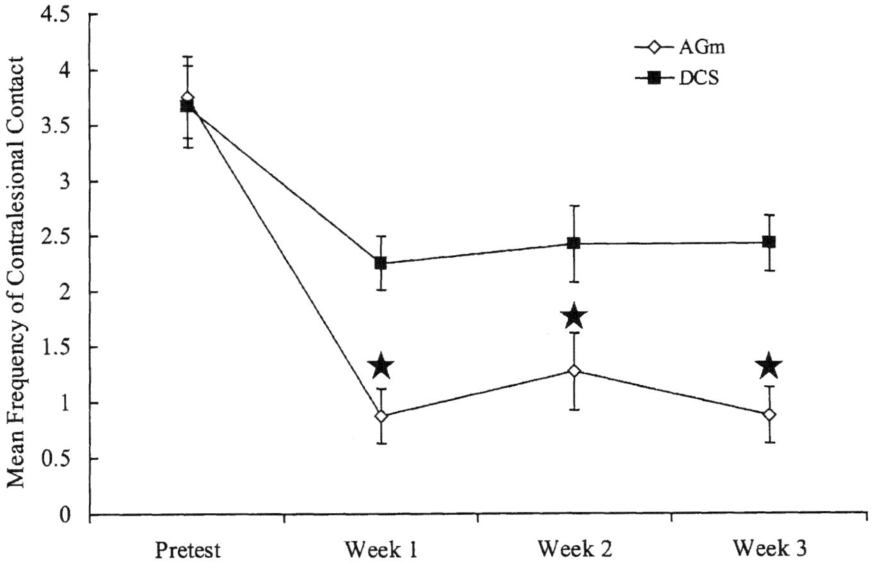

Figure 10.4. DCS-lesioned rats exhibit no extinction, in contrast to AGm-lesioned rats. Shown are mean frequencies of contacting the contralateral tab first in the AGm and DCS groups. Maximum number of potential removals is five. Stars indicate significant difference ($P < 0.05$) from pretest.

of the contexts in which behavioral recovery has been obtained in the rodent model.

The dissociation between neglect and extinction in these studies is consistent with recent findings in the literature on human neglect patients [Stone et al., 1998, Vallar et al., 1994]. The results strongly suggest that different pharmacological and environmental interventions may be required to promote recovery for the constellation of deficits that comprise the neglect syndrome. These findings are particularly important because in many “recovered” neglect patients extinction deficits remain [Kerkhoff, 2001]. These results have clear clinical implications, and they indicate that adequate treatment of patients with neglect and extinction may be pharmacologically complex, and that specific components of the "neglect syndrome" may have to be specifically addressed. This very issue was one of the highlights of a recent review of the literature on neglect [Kerkhoff, 2001].

2.2. Anatomical Studies

The anatomical connectivity of DCS has been examined in several studies by our group, beginning with the finding that DCS receives a prominent input from cortical area AGm (Figure 10.5). The anatomical studies have provided a foundation for quantitative assessment of dynamic changes in the DCS that are correlated with behavioral recovery.

Cortical projections to the DCS were investigated by injection of retrograde fluorescent axonal tracers into the DCS [Cheatwood et al., 2003]. A key finding of this study was that in addition to its main input from cortical area AGm, DCS also receives substantial input from the multimodal posterior parietal cortex (PPC). This is significant because PPC and AGm are linked by corticocortical connections [Vandevelde et al., 1996] and are both critical components of the circuitry involved in spatial processing and directed attention. Other cortical areas providing input to DCS include visual association area Oc2M, lateral agranular cortex (AGl), and orbital cortex (VLO, LO). These areas have reciprocal connections with AGm and PPC. Inconsistent labeling was seen in somatic sensorimotor areas FL, HL and Par 1.

Thalamic afferents to DCS were found to be prominent from LD, LP, MD, VL and the intralaminar nuclei. Collectively, these nuclei constitute the sources of thalamic input to cortical areas AGm and PPC. We found evidence of topography in the thalamic projections to DCS from LD and LP, and the observed pattern is consistent with the known organization of thalamocortical projections involving these nuclei and the topography of corticostriatal projections from their cortical targets. Injections in dorsal DCS consistently labeled thalamic nuclei LD and LP. Together with previous data on thalamocortical and corticocortical connections, this finding suggests that dorsal DCS is preferentially related to caudal AGm.

In a subsequent study [Reep et al., 2003] we utilized the anterograde axonal tracer BDA to delineate the pattern of corticostriatal terminations in DCS originating from areas AGm, PPC, Oc2M, AGl, and the orbital cortex. These findings revealed that the projection from AGm is prominent within DCS and the main corticostriatal projections from areas other than AGm are situated around the periphery of DCS in a roughly circular fashion: visual association cortex dorsomedially, PPC dorsally, AGl laterally, and orbital cortex ventrally (Figure 10.6). Each of these cortical projections is also represented by less dense aggregates of terminal labeling within DCS. Double anterograde labeling using fluorescent dextran tracers provided direct confirmation that corticostriatal axons from AGm and PPC (and from AGm and Oc2M) overlap and interdigitate at foci within DCS, often forming terminal fields within close proximity to one another. This relationship suggests that axons from AGm and PPC may be making synaptic connections on the same individual striatal neurons. We are now investigating this possibility using anterograde tracing coupled with electron microscopy.

An important discovery from the anatomical experiments was that DCS receives input from several cortical areas which are themselves interconnected. This indicates a selective convergence within DCS of projections from inter-

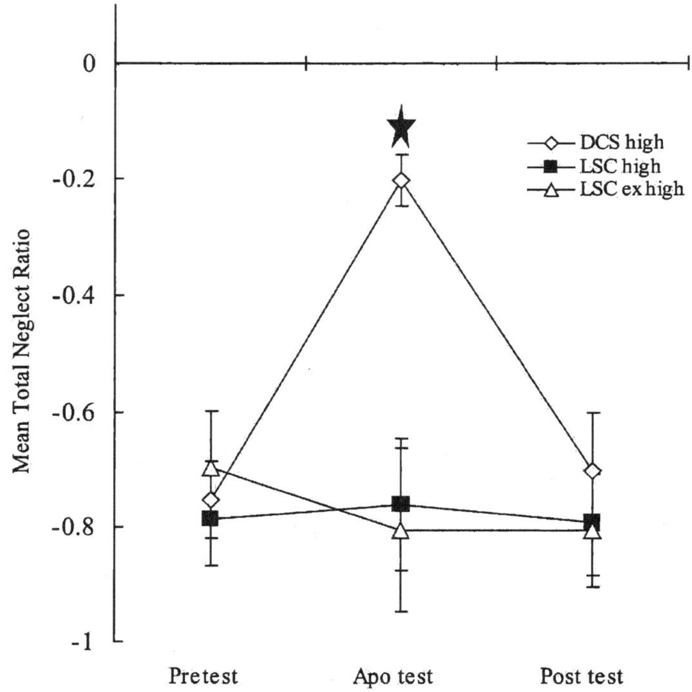

Figure 10.5. Neglect ratios for four groups with unilateral lesions of AGm, given different dosages of apomorphine or vehicle infused into DCS. A ratio of 0.0 represents symmetrical responding, and all groups exhibited severe neglect in the pretest phase. The star indicates that the high dosage group exhibited significantly more symmetrical responding than all other groups except the medium dosage group.

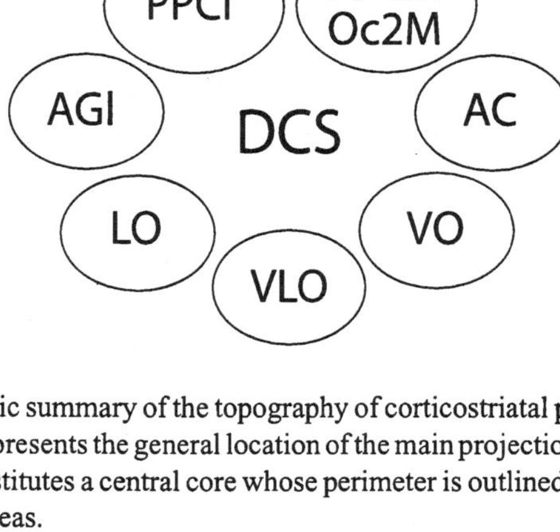

Figure 10.6. Schematic summary of the topography of corticostriatal projections in the vicinity of DCS. Each ellipse represents the general location of the main projection field from the indicated cortical area. DCS constitutes a central core whose perimeter is outlined by the dense projections from several cortical areas.

connected cortical areas, and represents support for the contention that DCS is a core associative region of the dorsal striatum. An especially interesting example is that projections from AGm and PPC converge in DCS. These two cortical areas are linked by extensive corticocortical connections [Corwin and Reep, 1998]. When these axons are selectively severed without directly damaging PPC or AGm, neglect occurs [Burcham et al., 1997]. Taken together with the behavioral and pharmacological data above, it appears that correlated convergent input to DCS from AGm and PPC may be necessary for normal directed attention and neglect.

The results of the anatomical studies to date indicate that DCS receives inputs from cortical and thalamic areas that are themselves linked by corticocortical and thalamocortical connections (Figure 10.7). Together, the findings of the behavioral and anatomical studies support the view that DCS is an integral component of an associative network of cortical, striatal, and thalamic regions involved in directed attention, is the site of multimodal integration of spatial stimuli, and is the critical substrate for recovery from neglect produced by cortical lesions.

3. Role of induced plasticity in DCS in Recovery from AGm-induced Neglect

In a number of recent studies the search for factors to promote plasticity and recovery of function from damage to the CNS has focused on the role of extrinsic factors in the neural environment that might be inhibitory for neurite outgrowth and plasticity [Brittis and Flannagan, 2001, Horner and Gage, 2000]. The results of these studies have indicated that plasticity can be induced following CNS injury by inhibition of the glial inhibitory molecule Nogo through the use of treatment with anti-Nogo antibodies such as IN-1 [Bandtlow and Schwab, 2000, Chen et al., 2000], and the identification of receptors that can mediate Nogo activity [Fournier et al., 2001, Liu et al., 2002]. IN-1 treatment has been found to produce recovery and plasticity in several contexts including regeneration of corticospinal tracts in cats [Schnell and Schwab, 1990, Schnell and Schwab, 1993, Schnell et al., 1994], functional recovery of locomotor functions [Bregman et al., 1995], and in a forelimb reaching task [Papadopoulos et al., 2002].

Our behavioral findings indicate that the striatal projection zone of the AGm in the dorsocentral striatum (DCS) is crucial for recovery of function from severe neglect following unilateral destruction of the AGm. It remains to be determined whether dynamic changes in the DCS are systematically related to recovery from AGm-induced neglect. In order to examine this issue we have embarked on a series of studies designed to determined whether IN-1 will induce plasticity in the DCS which would lead to subsequent behavioral recovery.

Table 10.1. BDA-labeled axon density in DCS at the septal level ($\sim +0.7$ ap) resulting from injections of BDA in cortical area AGm on the right side.

	Normal AGm	*Control hybridoma (left AGm lesioned)*	*IN-1 hybridoma (left AGm lesioned)*
Case numbers	108, 109, 112	H300, 355, 364	H302, 346, 347
Contra/Ipsi Axon density ratio	0.33–0.42	0.35–0.47	0.85–1.31

In our initial study, we examined whether administration of IN-1 would induce plasticity in the DCS following unilateral lesion of cortical area AGm. Given our prior findings concerning the crucial role of the DCS in recovery from neglect, we felt that such plasticity would have positive implications for recovery from neglect and the potential role of the input from contralesional AGm and ipsilesional PPC in recovery. Therefore, we examined whether IN-1 can induce neural growth in the afferents to the DCS from the contralesional AGm and the ipsilesional PPC, following unilateral lesion of AGm, and whether this plasticity is systematically related to recovery from severe AGm-induced neglect.

Prior studies of corticostriatal plasticity have focused on the input from the contralesional cortex [Carmichael and Chesselet, 2002, Kartje et al., 1999, Papadopoulos et al., 2002]. The monoclonal antibody IN-1 successfully induces sprouting of these projections in the form of increased axon density, and this plasticity is associated with functional recovery [Kartje et al., 1999, Papadopoulos et al., 2002]. However, sprouting may also be considered to include growth cone formation and synaptogenesis. For the AGm, the ipsilateral projection to the DCS in normal animals is approximately three times denser than the contralateral (Table 10.1). When the left AGm is lesioned to produce neglect, significant denervation of left DCS occurs. Does the input from right AGm to left DCS then sprout to occupy this vacated synaptic space (Figure 10.8)? If so, there should be an associated increase in axon density from the contralateral AGm. This change would be reflected as an increase in the contralateral/ipsilateral (left/right) ratio of axon density in DCS originating from the right AGm. This can be assessed by depositing an anterograde axonal tracer in right AGm and measuring axon density in left and right DCS [Papadopoulos et al., 2002].

3.1. Sprouting from contralesional AGm

A series of subjects (N=9) received lesions of left AGm which were accompanied by injection of 5μl of suspended hybridoma cells producing either the

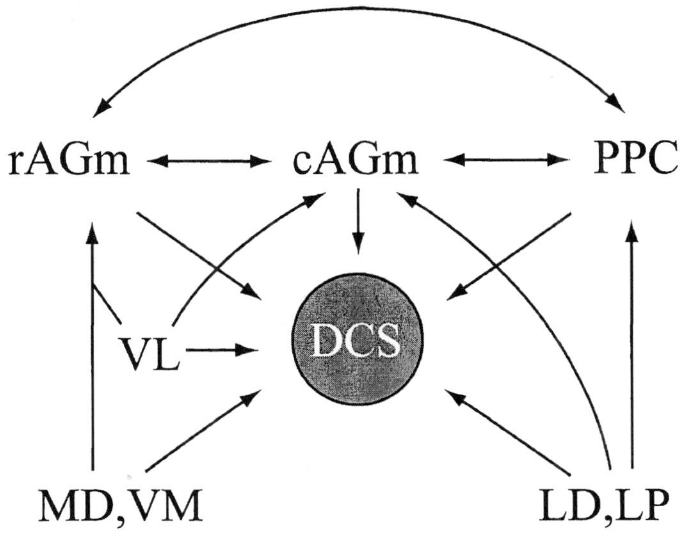

Figure 10.7. Major pattern of connections involving DCS. Thalamic regions projecting to DCS also project to cortical areas AGm and PPC, which converge in DCS.

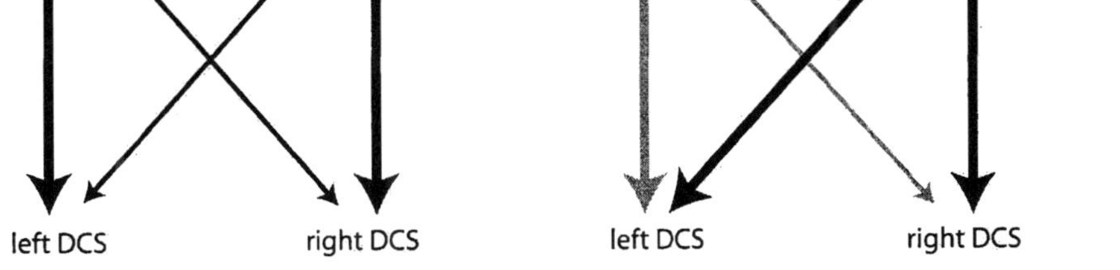

Figure 10.8. Schematic representation of the normal corticostriatal projections from AGm to dorsocentral striatum, and the hypothesized sprouting that occurs following unilateral lesion of AGm and IN-1 treatment.

IN-1 antibody or a control antibody (HRP). The subjects were tested for neglect for 4-7 weeks, and we then injected BDA in the right (contralesional) AGm and assessed sprouting by measuring axon density in DCS. In three normal control animals with injections of BDA in rostral, mid or caudal AGm, the axon density contralateral to the injection was less than half the density ipsilateral to the injection, as reflected in contralateral/ipsilateral ratios less than 0.5 (Table 1; Figure 10.9). Control hybridoma cases exhibited similar ratios. However, in IN-1 experimental cases that had recovered from severe neglect, the contralateral density increased by over 100%, resulting in densities equivalent to those ipsilaterally, and thus ratios of 1.0 (Table 1; see Figure 10.2). This preliminary evidence indicates that subjects that were given IN-1 demonstrated behavioral recovery and axonal sprouting in the left DCS from axons originating in contralesional AGm. Thus, the data suggest that IN-1 induces axonal plasticity in treated subjects, and that this plasticity may be the cause of the recovery from severe neglect. We found that the changes in density were not based on shrinkage of the DCS subsequent to the AGm lesion, which could have resulted in an apparent increase in density. Neither the DCS nor the striatum as a whole exhibited any shrinkage. These findings support those of Papadopoulos et al. [Papadopoulos et al., 2002] and Kartje et al. [Kartje et al., 1999] using IN-1, and Carmichael and Chesselet [Carmichael and Chesselet, 2002] using thermocoagulatory lesions, which demonstrate corticostriatal plasticity from the contralesional motor cortex. Our results suggest that those findings extend to corticostriatal inputs from association cortex, and that plasticity following IN-1 treatment may produce recovery from severe neglect. Moreover, the plasticity appears to be specific to areas that project to the DCS. In one case, after a lesion of left AGm, BDA was injected into the right anterior cingulate cortex instead of right AGm. This case exhibited no increase in axon density in left DCS compared to a control (contra/ipsi ratios = 0.30 and 0.35), suggesting that not all cortical areas projecting to DCS are competent to sprout into it.

In previous studies that have examined sprouting and its relationship to motor recovery, typically only contralesional inputs have been examined [Kartje et al., 1999, Papadopoulos et al., 2002]. In the case of neglect we have demonstrated that the AGm and the ipsilesional PPC function as a system [Burcham et al., 1998]. Destruction of either area alone or transection of connections between the two produce neglect. Given these relationships, we examined the potential for sprouting of the PPC projections to the DCS in two subjects. In an experiment similar to those described above, BDA was injected into the left PPC following lesion of the left AGm and subsequent treatment with IN-1. Normally, the projection from PPC to ipsilateral DCS is less than twice the density of the contralateral projection (ipsi/contra ratio = 1.78). However, after IN-1 treatment there was a large (60%) increase in this ratio, to 2.86. The change in density was not explainable by shrinkage of the DCS which could lead to an

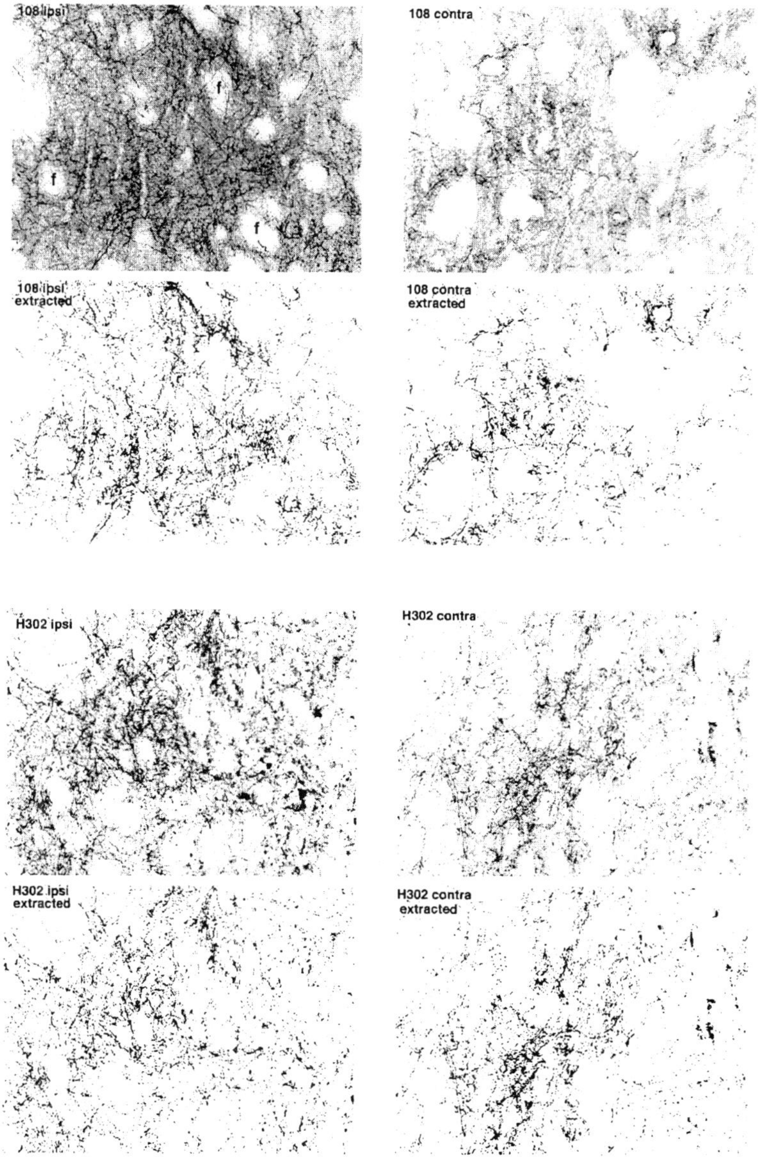

Figure 10.9. All images depict BDA-labeled axons in the dorsocentral striatum (DCS). For each brain (cases 108 and H302) four panels are shown, depicting ipsilateral and contralateral axonal labeling after an injection of BDA into cortical area AGm. Fascicles are denoted by "f". For each brain the top row represents normalized, contrast-enhanced images, and the bottom row represents extracted (high pass filtered) versions of the same images. The extracted images were used for density measurements. In normal rats like case 108 there is denser axonal labeling in the ipsilateral DCS than contralaterally. In experimental cases like H302 there was a shift so that the contralateral side exhibited axon density equivalent to that on the ipsilateral side.

apparent change in density. These preliminary findings are strongly suggestive that sprouting occurred in the PPC projections into the DCS.

The results of these preliminary studies strongly suggest that robust sprouting effects can be produced in this system in the presence of the IN-1 antibody. Further, these results suggest that cortical inputs to DCS other than the crossed input from AGm may also exhibit sprouting. A related issue which may speak to the potential clinical efficacy of IN-1concerns the specificity of the plasticity. We have found plasticity in the ipsi and contralesional inputs to the DCS. Is the plasticity specific to DCS or is it also found in other regions of the dorsal striatum? Perhaps IN-1 treatment produces uncontrolled nonspecific plasticity in the striatum. We have preliminary data from one subject in which the subject received an AGm lesion and was treated with IN-1. After behavioral recovery BDA was injected into the contralesional cingulate cortex which projects to an adjacent region of striatum, but not to the DCS. The results indicated that the ratio of ipsi/contralateral inputs from the cingulate cortex into the striatum is unchanged despite the presence of IN-1. Also, as mentioned above, there was no evidence of sprouting from the cingulate cortex into the DCS even though the main striatal projection zone of the cingulate cortex is medially adjacent to the DCS.

4. Clinical Significance

Neglect is a severe and prevalent clinical disorder, and at present there are no generally accepted therapies for the treatment of neglect. We have developed a rodent model to study neglect and behavioral recovery, and these rats exhibit many of the same fundamental types of deficits found in human patients with neglect. The evidence discussed above, including our preliminary findings, strongly suggests that the DCS plays a crucial role in neglect induced by cortical lesions, and that DCS may be the crucial site for the mechanisms leading to recovery. Our anatomical findings indicate that the DCS receives converging inputs from the AGm and the PPC, and that the DCS might correctly be considered to be associative striatum [Cheatwood et al., 2003, Reep et al., 2003]. These anatomical findings have been the foundation for our most dramatic recent finding that extensive plasticity can be induced in the DCS afferents from the contralesional AGm and the ipsilesional PPC using IN-1, and that recovery from neglect is correlated with this plasticity. These findings suggest that induced axonal sprouting and synaptogenesis in the DCS may underlie recovery. Prior studies using IN-1 have focused on recovery from motor deficits [Papadopoulos et al., 2002]. In the present proposed studies we extend the potential use of growth-inducing factors into the treatment of neglect, a cognitive deficit induced by brain damage. What is most exciting and the essence of these experiments is that if plasticity can be induced there might eventually be a significant lasting treatment or a cure for this cognitive disorder in humans.

Acknowledgments

This research was supported by NIMH grant MH60399. We greatly appreciate the technical assistance of Scott Heldt, Michael Hylin, Maggie Stoll, and Steve Wagner, and the imaging work of Gabriel Trinity.

References

[Bandtlow and Schwab, 2000] Bandtlow, C. E. and Schwab, M. E. (2000). NI-35/250/nogo-a:a neurite growth inhibitor restricting structural plasticity and regeneration of nerve fibers in the adult vertebrate CNS. *Glia*, 29:175–181.

[Bregman et al., 1995] Bregman, B. S., Kunkel-Bagden, E., Schnell, L., Dai, H. N., Gao, D., and Schwab, M. E. (1995). Recovery from spinal cord injury mediated by antibodies to neurite growth inhibitors. *Nature*, 378:498–501.

[Brittis and Flannagan, 2001] Brittis, P. A. and Flannagan, J. G. (2001). Nogo domains and a Nogo receptor: implications for axon regeneration. *Neuron*, 30:11–14.

[Burcham and Corwin, 1998] Burcham, K. J. and Corwin, J. V. (1998). Effects of delay and duration of light deprivation on recovery of function from neglect induced by unilateral medial agranular prefrontal cortex lesions in rats. *Psychobiol.*, 26:216–260.

[Burcham et al., 1997] Burcham, K. J., Corwin, J. V., Stoll, M. L., and Reep, R. L. (1997). Disconnection of medial agranular and posterior parietal cortex produces multimodal neglect in rats. *Behav. Brain Res.*, 86:41–47.

[Burcham et al., 1998] Burcham, K. J., Corwin, J. V., and Van Vleet, T. (1998). Light deprivation produces behavioral recovery of function from multimodal neglect following unilateral posterial parietal cortex lesions in rats. *Behav. Brain Res.*, 90:187–197.

[Caplan et al., 1990] Caplan, L. R., Schmahmann, J. D., Kase, C. S., Feldmann, E., Baquis, G., Greenberg, J. P., Gorelick, P. B., Helgason, C., and Hier, D. B. (1990). Caudate infarcts. *Arch. Neurol.*, 47:133–143.

[Carmichael and Chesselet, 2002] Carmichael, S. T. and Chesselet, M. F. (2002). Synchronous neuronal activity is a signal for axonal sprouting after cortical lesions in the adult. *J. Neurosci.*, 22:6062–6070.

[Cheatwood et al., 2003] Cheatwood, J., Reep, R., and Corwin, J. (2003). The associative striatum: cortical and thalamic projections to the dorsocentral striatum. *Brain Res.*, 968:1–14.

[Chen et al., 2000] Chen, M. S., Huber, A. B., Van der Haar, M. E., Frank, M., Schnell, L., Spillmann, A. A., Christ, F., and Schwab, M. E. (2000). Nogo-A is a myelin-associated neurite outgrowth inhibitor and an antigen for monoclonal antibody IN-1. *Nature*, 403:434–439.

[Chen and Hillman, 1990] Chen, S. and Hillman, D. E. (1990). Robust synaptic plasticity of striatal cells following partial deafferentation. *Brain Res.*, 520:103–114.

[Corwin et al., 1986] Corwin, J. V., Kanter, S., Watson, R. T., Heilman, K. M., Valenstein, E., and Hashimoto, A. (1986). Apomorphine has a therapeutic effect on neglect produced by unilateral dorsomedial prefrontal cortex lesions in rats. *Exper. Neurol.*, 94:683–689.

[Corwin and Reep, 1998] Corwin, J. V. and Reep, R. L. (1998). Rodent posterior parietal cortex as a component of a cortical network mediating directed spatial attention. *Psychobiol.*, 26:87–102.

[Couvier et al., 1987] Couvier, W., Bua, B., Blanton, P., and Urey, J. (1987). Behavioral changes following visual scanning training: observation of five cases. *Int. J. Clin. Neuropsych.*, 9:74–80.

[Crowne et al., 1983] Crowne, D. P., Richardson, C. M., and Ward, G. (1983). Brief deprivation of vision after unilateral lesions of the frontal eye field prevents contralateral inattention. *Science*, 220:527–530.

[Denes et al., 1982] Denes, G., Semenza, C., Stoppa, E., and Lis, A. (1982). Unilateral spatial neglect and recovery from hemiplegia: A follow-up study. *Brain*, 105:543–552.

[Fleet et al., 1987] Fleet, W. S., Valenstein, E., Watson, R. T., and Heilman, K. M. (1987). Dopamine agonist therapy for neglect in humans. *Neurology*, 37:1765–1770.

[Fournier et al., 2001] Fournier, A. E., GrandPre, T., and Strittmatter, S. M. (2001). Identification of a receptor mediating Nogo-66 inhibition of axonal regeneration. *Nature*, 409:341–346.

[Fullerton et al., 1997] Fullerton, K. J., Mackenzie, G., and Stout, R. W. (1997). Prognostic indices in stroke. *Quart. J. Med.*, 66:147–162.

[Geminiani et al., 1998] Geminiani, G., Bottini, G., and Sterzi, R. (1998). Dopamine stimulation in unilateral neglect. *J. Neurosurg. Psychiat.*, 65:345–347.

[Halligan and Marshall, 1994] Halligan, P. W. and Marshall, J. C. (1994). Current issues in spatial neglect: an editorial introduction. *Neuropsych. Rehab.*, 4:103–110.

[Heilman et al., 1993] Heilman, K. M., Watson, R. T., and Valenstein, E. (1993). Neglect and related disorders. In Heilman, K. M. and Valenstein,

E., editors, *Clinical Neuropsychology*, pages 279–336. Oxford University Press, New York, NY, third edition.

[Henley et al., 1985] Henley, S., Pettit, P., Todd-Pokropek, L., and Tupper, L. (1985). Who goes home? Predictive factors in stroke recovery. *J. Neurol. Neurosurg. Psychiat.*, 48:1–6.

[Horner and Gage, 2000] Horner, P. J. and Gage, F. H. (2000). Regenerating the damaged central nervous system. *Nature*, 407:963–970.

[Hurford et al., 1998] Hurford, P., Stringer, A. Y., and Jann, B. (1998). Neuropharmacologic treatment of hemineglect: A case report comparing bromocriptine and methylphenidate. *Arch. Physical Med. Rehab.*, 79:346–349.

[Karnath et al., 2001] Karnath, H., Ferber, S., and Himmelbach, M. (2001). Spatial awareness is a function of the temporal not the posterior parietal lobe. *Nature*, 411:950–953.

[Kartje et al., 1999] Kartje, G. L., Schulz, M. K., Lopez-Yunez, A., Schnell, L., and Schwab, M. E. (1999). Corticostriatal plasticity is restricted by myelin-associated neurite growth inhibitors in the adult rat. *Ann. Neurol.*, 45:778–786.

[Kerkhoff, 2001] Kerkhoff, G. (2001). Spatial hemineglect in humans. *Prog. Neurobiol.*, 63:1–27.

[King and Corwin, 1990] King, V. and Corwin, J. V. (1990). Neglect following unilateral ablation of the caudal but not the rostral portion of the medial agranular cortex of the rat. *Behav. Brain Res.*, 37:169–184.

[King and Corwin, 1992] King, V. and Corwin, J. V. (1992). Spatial deficits and hemispheric asymmetries in the rat following unilateral and bilateral lesions of the posterior parietal or medial agranular cortices. *Behav. Brain Res.*, 50:53–68.

[King and Corwin, 1993] King, V. and Corwin, J. V. (1993). Comparisons of hemiinattention produced by unilateral lesions of the posterior parietal or the medial agranular prefrontal cortex in the rat. *Behav. Brain Res.*, 54:117–131.

[Liu et al., 2002] Liu, B. P., Fournier, A., GrandPre, T., and Strittmatter, S. M. (2002). Myelin-associated glycoprotein as a functional ligand for the Nogo-66 inhibition of axonal regeneration. *Science*, 297:1190–1193.

[Marshall and Gotthelf, 1979] Marshall, J. F. and Gotthelf, T. (1979). Sensory inattention in rats with 6-hydroxdopamine-induced degeneration of ascending dopaminergic neurons: Apomorphine-induced reversal of deficits. *Exper. Neurol.*, 65:398–411.

[Mesulam, 1980] Mesulam, M. M. (1980). A cortical network for directed attention and unilateral neglect. *Ann. Neurol.*, 10:309–325.

[Mesulam, 1990] Mesulam, M. M. (1990). Large-scale neurocognitive networks and distributed processing for attention, language, and memory. *Ann. Neurol.*, 28:595–613.

[Milner, 1987] Milner, A. D. (1987). Animal models for the syndrome of spatial neglect. In Jeannerod, M., editor, *Neurophysiological and neuropsychological aspects of spatial neglect*, pages 259–288. North-Holland, Amsterdam.

[Papadopoulos et al., 2002] Papadopoulos, C. M., Tsai, S. Y., Alsbiei, T., O'Brien, T. E., Schwab, M. E., and Kartje, G. L. (2002). Functional recovery and neuroanatomical plasticity following middle cerebral artery occlusion and IN-1 antibody treatment in the adult rat. *Ann. Neurol.*, 51:433–441.

[Parent, 1990] Parent, A. (1990). New frontiers in basal ganglia research. *Trends Neurosci.*, 13:241–244.

[Pyter et al., 2002] Pyter, B., Heldt, S. A., Van Vleet, T. M., Corwin, J. V., and Reep, R. L. (2002). Apomorphine produces recovery of function from neglect, but not extinction deficits in rats with unilateral medial agranular cortex lesions. *Soc. Neurosci. Abst.*, 28.

[Reep et al., 2003] Reep, R. L., Cheatwood, J. L., and Corwin, J. V. (2003). The associative striatum: organization of cortical projections to the dorsocentral striatum in rats. *J. Comp. Neurol.*, (In press).

[Reep and Corwin, 1999] Reep, R. L. and Corwin, J. V. (1999). Topographic organization of the striatal and thalamic connections of rat medial agranular cortex. *Brain Res.*, 841:43–52.

[Robertson et al., 1993] Robertson, I. H., Halligan, P. W., and Marshall, J. C. (1993). Prospects for the rehabilitation of unilateral neglect. In Robertson, I. and Marshall, J. C., editors, *Unilateral Neglect: Clinical and Experimental Studies*, pages 279–292, Hillsdale, NJ. Lawrence Erlbaum Associates.

[Schnell et al., 1994] Schnell, L., Schneider, R., Kolbeck, R., Barde, Y. A., and Schwab, M. E. (1994). Neurotrophin-3 enhances sprouting of corticospinal tract during development and after adult spinal cord lesion. *Nature*, 367:170–173.

[Schnell and Schwab, 1990] Schnell, L. and Schwab, M. E. (1990). Axonal regeneration in the rat spinal cord produced by an antibody against myelin-associated neurite growth inhibitors. *Nature*, 343:269–272.

[Schnell and Schwab, 1993] Schnell, L. and Schwab, M. E. (1993). Sprouting and regeneration of lesioned corticospinal tract fibres in the adult rat spinal cord. *Euro. J. Neurosci.*, 5:1156–1171.

[Stone et al., 1998] Stone, S. P., Halligan, P. W., Marshall, J. C., and Greenwood, R. J. (1998). Unilateral neglect: A common but heterogeneous syndrome. *Neurology*, 50:1902–1905.

[Vallar et al., 1994] Vallar, G., Rusconi, M. L., Bignamini, L., Geminiani, G., and Perani, D. (1994). Anatomical correlates of visual and tactile extinction in humans: a clinical CT scan study. *J. Neurol. Neurosurg. Psychiat.*, 57:464–470.

[Van Vleet et al., 2000] Van Vleet, T. M., Burcham, K. J., Corwin, J. V., and Reep, R. L. (2000). Unilateral destruction of the medial agranular cortical projection zone in the dorsocentral striatum produces severe neglect in rats. *Psychobiol.*, 28:57–66.

[Van Vleet et al., 2003a] Van Vleet, T. M., Corwin, J. V., Heldt, S. A., Pyter, B., and Reep, R. L. (2003a). Effects of light deprivation on recovery from neglect and extinction induced by unilateral lesions of the medial agranular cortex and dorsocentral striatum. *Behav. Brain Res.*, 138:165–178.

[Van Vleet et al., 2002] Van Vleet, T. M., Guerrettaz, K., Heldt, S. A., Corwin, J. V., and Reep, R. L. (2002). Unilateral destruction of the dorsocentral striatum in rats produces neglect but not extinction to bilateral simultaneous stimulation. *Behav. Brain Res.*, 136:375–387.

[Van Vleet et al., 2003b] Van Vleet, T. M., Heldt, S. A., Corwin, J. V., and Reep, R. L. (2003b). Infusion of apomorphine into the dorsocentral striatum produces acute recovery from neglect induced by unilateral medial agranular cortex lesions in rats. *Behav. Brain Res.*, 143:147–157.

[Vandevelde et al., 1996] Vandevelde, I., Duckworth, E., and Reep, R. L. (1996). Layer VII and the gray matter trajectory of corticocortical axons in rats. *Anat. Embryol.*, 194:581–593.

[Vargo et al., 1989] Vargo, J. M., Richard-Smith, M., and Corwin, J. V. (1989). Spiroperidol reinstates asymmetries in neglect in rats recovered from left versus right dorsomedial prefrontal cortex lesions. *Behav. Neurosci.*, 103:1017–1027.

Chapter 11

BINARY AND SPARSE CHECKERBOARD VISUAL STIMULI IN MULTIPLE SCLEROSIS PATIENTS

Rasa Ruseckaite

Teddy Maddess

Andrew Charles James

Centre for Visual Sciences
Research School of Biological Sciences
Australian National University
Canberra, ACT 0200
Australia
rasa.ruseckaite@anu.edu.au

Introduction

Visual evoked potentials (VEPs) to checkerboard stimuli are often recorded to evaluate the functional visual nervous system from the eye to brain. For example, various types of VEPs have been used for the early detection of multiple sclerosis (MS) [Harter, 1970, Chiappa, 1983, Andersson and Siden, 1991, Towle et al., 1991, Hood and Zhang, 2000]. MS is a central nervous system (CNS) disease, characterized by multiple areas of demyelination [Robinson and Rudge, 1977, Waxman, 1983]. People suffering from MS often develop optic neuritis (ON) as well.

We recorded the multifocal VEPs (mVEPs) from the Normal and MS subjects using multifocal stimuli, having four degrees of temporal sparseness: Binary, $Sparse_4$, $Sparse_{16}$ and Pattern Pulse [James, 2003]. We examined response waveforms and latencies at different densities types of visual stimuli to find out which one produced the significant responses and the largest delays in MS subjects.

1. Methods

1.1. Stimuli

The description of VEP stimuli is given in our earlier study [Maddess et al., 2003]. Subjects viewed the monitor from a distance of 30 cm providing the stimulus layout as illustrated in Figure 11.1. The monitor was divided into eight regions, and red fixation spot was presented at the screen's centre.

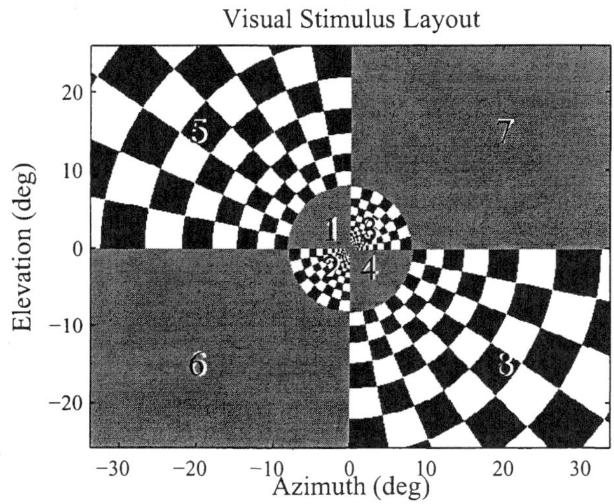

Figure 11.1. Sample of the visual stimulus. In practice the checks had contrasts 1 or 0 (i.e. black, white or grey). The numbers (1,. . .,8) indicate the eight different regions.

The examples of four types of temporal sequences are presented in Figure 11.2.

1.2. Subjects

The MS study group contained 50 subjects (eight men and 42 women, age range 25 to 64 (45 ± 15.2 year). 26 subjects suffered from ON. The Normal study group contained 19 subjects (12 men and seven women, age range 22 to 44 (31.2 ± 13.1 year), with normal or corrected to normal refraction). The summarized subjects' data are presented in Table 11.1.

The research followed the tenets of the Declaration of Helsinki, under the Australian National University's Human Experimentation Ethics Committee under protocol M9901. Informed written consent was obtained from the subjects after the nature and possible consequences of the study were explained to them.

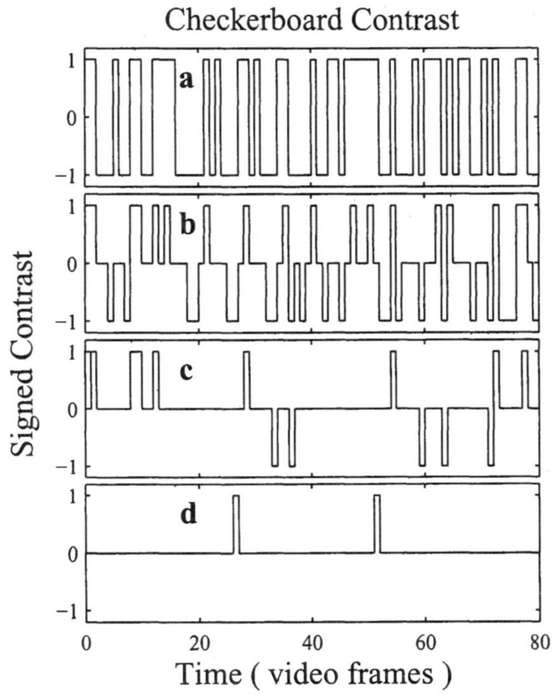

Figure 11.2. Example of the temporal modulation of a single region. a) Temporal stimulus density for the first Binary sparseness (probability of $\frac{1}{2}$ that check contrast takes the values -1 or 1). b) Sparse 4 stimulus, when the probability of a checkerboard appears in one sign of the other was $\frac{1}{4}$; the check contrast takes the values $\{-1, 0, 1\}$). c) Sparse 16 when the probability of a checkerboard appears in one sign of the other is $\frac{1}{16}$. The stimulus is ternary; the check contrast taking the values $\{-1, 0, 1\}$. d) Pattern Pulse, when the probability of a checkerboard appears in one sign of the other is $\frac{1}{39}$. The stimulus is ternary; check contrast taking the values $\{-1, 0, 1\}$. The duration of one video frame (*Time* axis) is 19.7 ms/eye.

Table 11.1. Summarized subject data. The two columns at the left show two study groups (MS and Normals) and a number of subjects. VEPs were recorded for 8 MS and 12 Normal men (see Sex). Duration of MS indicates the age of the disease. During this time patients had approximately 9.46 clinical attacks (see N of attacks). 26 subjects had optical neuritis (ON column) and for 16 of them CSF test was positive (CSF). MS type was Relapsing Remitting (RR).

Study group	*N*	*Age* ± *SE (yr)*	*Sex (M/F)*	*Duration of MS (yr)*	*N of attacks*	*ON*	*CSF*	*MS type*
MS	50	45 ± 15.2	8/42	8.72 ± 7.2	9.46 ± 5.2	26	16	RR
Normals	19	31.2 ± 13.1	12/7	NA	NA	NA	NA	NA

2. Data Analysis

Within each response we analyzed positive and negative time peaks within two periods in Normal and MS responses. Normal study group contained the first two peaks of all the responses in [59.4 to 99 ms] and [100 to 158 ms] time periods respectively. Since these two time windows contained the first negativity (N1) and the first positivity (P1), the peaks and their temporal windows are called as the N1 and P1 peaks and windows [Maddess et al., 2003].

The first MS response negativities N1MS ad positivities P1MS were found in a whole length of response [40 to 300 ms]. To find them we used a special criterion $crit = 1.45$ multiplied by a standard error (SE) of the data. This criterion allowed us to estimate the best time peaks at a small error rate.

2.1. Multivariate regression analysis

Our goal was to find out whether the MS waveforms could be decomposed into several ones, and whether they would be delayed and how much. We used a multivariate regression model to find such a decomposition. Coefficients, obtained by means of the multivariate regression, described the fit between individual MS subject data and averaged Normal data sets. The best-fitted coefficients stated about the possible MS data delay in comparison to the averaged Normal data.

2.2. Discriminant data analysis

The objective of this analysis was to determine whether the structure of the data permitted a method that was able to discriminate normal subjects from those who suffered from MS. We employed two types of discriminant analysis: Linear (LDA) and Quadratic discriminant analysis (QDA) [Johnson and Wichern, 1992].

3. Results

3.1. General findings

We obtained four repeats of the four response data sets, for each of the 16-stimulus regions (eight per eye). The sample results are shown in the Figure 11.3.

The left panel of Figure 11.3 represents the typical Normal Pattern Pulse responses, and the right panel the ON Pattern Pulse typical responses.

We observed a significant difference between Normal and MS multifocal responses. To examine the difference between Normal and MS data we again applied multiple linear regression analysis. Table 11.2 summarizes regression results.

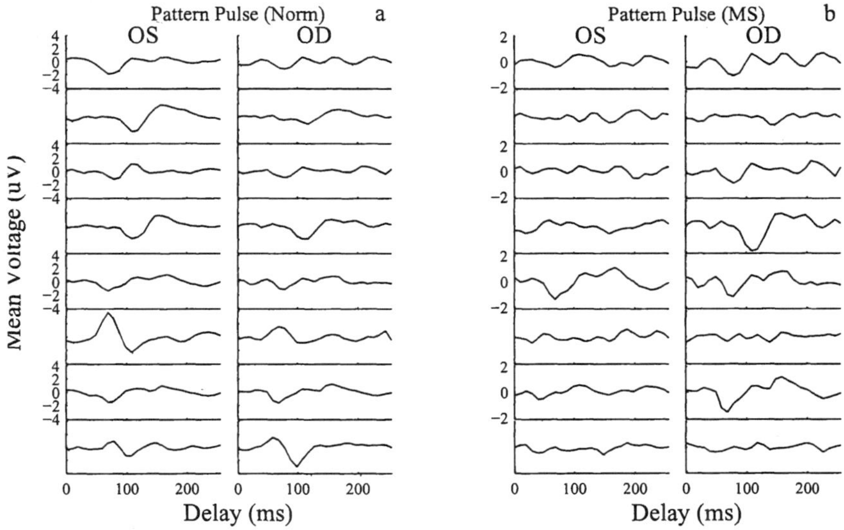

Figure 11.3. Typical Normals and MS responses. a) The two panels on the left (Pattern Pulse (Norm)) represent the multifocal responses for the left (OS) and for the right (OD) eye. b) The two right panels (Pattern Pulse (MS)) show the typical Pattern Pulse MS responses for the left (OS) and for the right (OD) eye, for the eight visual field regions (see Fig. 11.1). The eight rows of responses in each panel represent eight visual regions. Responses are shown in voltages (vertical axis). The horizontal axis indicates latencies in ms.

Table 11.2. Summarized multivariate linear regression results for MS vs Normals data. The fitted N1 peaks for SNR data are presented in dB. The column labelled *Condition* indicates fitted values. The reference condition was MS Binary. The Multiplier column represents the corresponding condition multiplicative factor. The conditions *$Sparse_4 * MS$*, *$Sparse_{16} * MS$* and *Pattern Pulse * MS* indicate interactions between $Sparse_4$, $Sparse_{16}$, Pattern Pulse and MS vs. Normals SNR N1 respectively. The coefficient for $Sparse_{16}$ stimulus is largest. According to this data we state, that MS data are 10.78 times smaller than Normal for $Sparse_{16}$ stimulus.

Condition	*Coefficient (dB)*	*SE (dB)*	*t*	*p*	*Multiplier* (×0.048)
Ref = Binary	−26.21	0.41	−30.9	0.0000	1
$Sparse_4 * MS$	0.92	0.96	1.76	0.0000	1.12
$Sparse_{16} * MS$	−10.78	0.53	−18.1	0.0000	0.28
Pattern Pulse * MS	−5.68	0.84	10.3	0.0000	1.92
Superior VF	−2.4	0.38	−10.5	0.0000	0.75

The data used in the present analysis were N1 absolute values of signal to noise ratios (SNR). The responses were converted do dB before fitting the model. Fitting simultaneously the Normal and MS study groups N1 interactions between each of sparseness, and a Superior visual field effect provided the most parsimonious model. For the maxima the variance accounted was $r^2 = 0.541$. The reference condition (-26.01 dB) corresponds to a Binary Normal data. The coefficients for Sparse_4 MS (0.92 dB), Sparse_{16} MS (-10.78 dB) and Pattern Pulse MS (-5.68 dB) corresponds to decreases of the responses for MS subjects group (compared with Normal data sets) from Binary figure by 1.12, 0.28 and 1.92 times respectively. We found that the MS data were significantly smaller than the Normal responses.

3.2. Delays

MS patients tended to have delayed VEPs. Previous investigations showed that VEPs latencies were especially prolonged for ON subjects [Hood et al., 2000b, Hood et al., 2000a]. In order to find out, whether our MS data had prolonged latencies or consisted of several waveforms, we applied the multivariate regression method, described in the Methods. The significance level of delay was determined by means of regression coefficient and p-level ($p < 0.05$).

The best-fitted delays T_1 were estimated for all MS subjects. Figure 11.4 illustrates the regression results for only 26 ON subjects. The picture represents the averaged delays (in video frames, $\sim$ 9.9 ms) for all stimulated regions and stimuli. All ON responses were delayed at least per one video frame. The largest delays (3 video frames) were obtained for Pattern Pulse visual stimuli.

We also estimated the second fitted delay T_2. We found that only 18 ON subjects had the second significant ($p < 0.05$) delay. Amongst those 18 subjects the second delay was found in regions 2, 4, 7 and 8 for the right eye. The averaged T_2 were: $\sim 3.6, 3.9, 3.0$ and 4.2 video frames for the regions $2, 4, 7$ and 8, for all of sparseness respectively.

At the next stage we compared the fitted delays together with the real MS data time to peaks (the estimation of $N1_{MS}$, NT_{MS}, $P1_{MS}$ and PT_{MS} is described in *Methods*). The NT_{MS} were considered. We estimated the longest delay T_F, which was picked from T_1 or T_2. I.e. if the data contained more than one-fitted delays, the longest one was chosen. The artificial data sets NT_F were created for all MS subjects, by adding together the longest fitted delays TF and Normal averaged first negativity time to peaks (NT). We expected to obtain an approximate MS time to peak values, which could be compared with the real one NT_{MS}. The fitted NT_F were subtracted from the NT_{MS} data sets. The differences ranged between -0.5 and 4 video frames, however they varied for different regions.

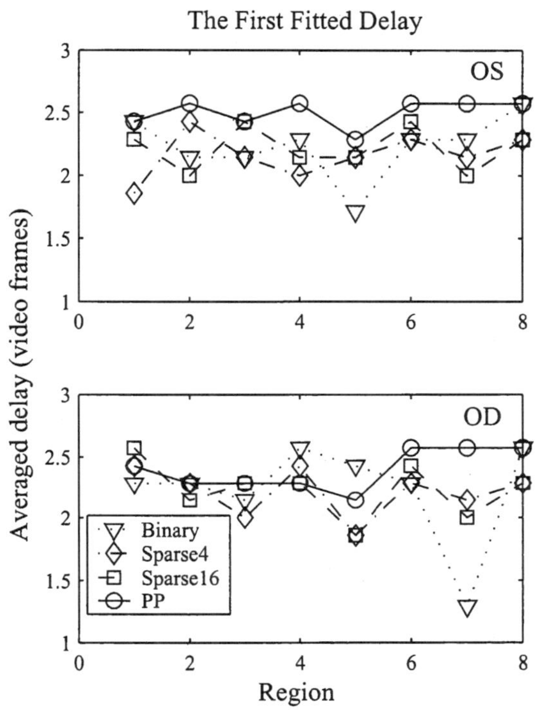

Figure 11.4. The first fitted delay T_1. The top panel (OS) represents the averaged T_1 for the left eye. The panel (OD) shows the T_1 for the right eye. The delays are averaged across all ON subjects, and are shown for each of temporal sparseness. The vertical axis shows the delay in video frames (1 video frame = 9.9 ms). The horizontal axis symbolizes regions 1 to 8 for each eye.

3.3. Discriminant data analysis

We next examined the specificity and sensitivity of ON VEP's responses. The first involved constructing models on the measures, obtained from each data group. We examined the effect of N1, P1 (see *Methods*), their implicit times, as well as the artificially obtained delays. Sensitivity and specificity was examined for each eye and region separately and together. One-eye responses were of little use.

The first model was constructed from N1, P1 and relevant time to peaks (NT and PT respectively) data sets. The LDA and QDA specificities and sensitivities were tested for each sparseness, eye and regions. The best specificities and sensitivities obtained were [91.3% 89.47%] for LDA and [94.7% 94.7%] for QDA.

We have also tested ON delays, obtained by using the multivariate regression (see *Methods*). We constructed artificial MS data sets by adding together averaged Normal data NT means and delayed components T_1 or T_2. A new discriminant model contained N1, NT, P1, PT values and one of those artificial data sets (NT_1 or NT_2) . The last combination was little of use in comparison with the NT_1. In this case the sensitivity and specificity increased till 100% for both LDA and QDA models.

4. Discussion

VEPs were assessed in four different levels of sparseness. To estimate possible delays in MS data we employed the multivariate regression model, which was described in *Methods*. We found that ON multifocal VEPs were delayed in $\sim$ 30ms than Normal responses. The delays varied between eyes. This happened because of a different level of optic neuritis. Our results showed bigger delays for sparser stimuli (Fig. 11.4). The delay varied for the different visual regions as well.

After applying the different discriminant models, we found high sensitivities and specificities for the model incorporating the P1, the N1, implicit times and the first fitted delay T_1. The best specificities and sensitivities were obtained again for the sparser stimuli. We also found it was worth combining the data of two eyes. In this case the performances reached [94.4% 94.7%] for LDA and [100% 100%] for QDA model.

Even though the simpler linear discriminant model performed at 100%, the more complex quadratic discriminant models require further verification. The latencies analysis showed that none of the regions had a very significant or different delay. Thus, fact that some regions give better performance can be explained as a random factor, dependent on nature of the data.

References

[Andersson and Siden, 1991] Andersson, T. and Siden, A. (1991). Multimodality evoked potentials and neurological phenomenology in patients with multiple sclerosis and potentially related conditions. *Electromyogr Clin Neurophysiol*, 31:109–117.

[Chiappa, 1983] Chiappa, K. (1983). *Evoked potentials in clinical medicine*, chapter Pattern shift visual evoked potentials: interpretation, pages 63–104. Raven Press, New York.

[Harter, 1970] Harter, M. (1970). Evoked cortical responses to checkerboard patterns: effect of check-size as a function of retinal eccentricity. *Vision Res*, 10:1365–1376.

[Hood et al., 2000a] Hood, D., Odel, J., and Zhang, X. (2000a). Tracking the recovery of local optic nerve function after optic neuritis: a multifocal VEP study. *Invest Ophthalmol Vis Sci*, 41:4032–4038.

[Hood and Zhang, 2000] Hood, D. and Zhang, X. (2000). Multifocal ERG and VEP responses and visual fields: comparing disease - related changes. *Doc Ophthalmol*, 100:115–137.

[Hood et al., 2000b] Hood, D., Zhang, X., Greenstein, V., Kangovi, S., Odel, J., Liebmann, M., and Ritch, R. (2000b). An interocular comparison of the multifocal VEP: a possible technique for detecting local damage to the optic nerve. *Invest Ophthalmol Vis Sci*, 41:1580–1587.

[James, 2003] James, A. (2003). The pattern pulse multifocal visual evoked potential. *Invest Ophthalmol Vis Sci*, (in press).

[Johnson and Wichern, 1992] Johnson, R. and Wichern, D. (1992). *Applied multivariate statistical analysis*. Prentice-Hall, Inc., third edition.

[Maddess et al., 2003] Maddess, T., Ruseckaite, R., and James, A. (2003). Effect of temporal sparseness and dichoptic presentation upon multifocal visual evoked potential. *J Neurosci*, (in press).

[Robinson and Rudge, 1977] Robinson, K. and Rudge, P. (1977). Abnormalities of the auditory evoked potentials in patients with multiple sclerosis. *Brain*, 100:19–40.

[Towle et al., 1991] Towle, V., Witt, J., Nader, S., Reder, A., Foust, R., and Spire, J. (1991). Three-dimensional human pattern visual evoked potentials. ii. multiple sclerosis patients. *Electroencephalogr Clin Neurophysiol*, 80:339–346.

[Waxman, 1983] Waxman, S. (1983). The demyelinating diseases. *Clinical Neurosc*, 1:609–643.

Chapter 12

SPATIOTEMPORAL TRANSITIONS IN TEMPORAL LOBE EPILEPSY*

J.C. Sackellares
Departments of Neuroscience, Neurology, Pediatrics, and Biomedical Engineering, University of Florida; VA Medical Center, Gainesville, FL
sackellares@mbi.ufl.edu

L.D. Iasemidis
Department of Biomedical Engineering, Arizona State University
Leon.Iasemidis@asu.edu

D.-S. Shiau
Department of Neuroscience, University of Florida; VA Medical Center, Gainesville, FL
shiau@epilepsy.health.ufl.edu

P.M. Pardalos
Departments of Industrial and Systems Engineering, Computer Science, and Biomedical Engineering, University of Florida
pardalos@cao.ise.ufl.edu

P.R. Carney
Department of Pediatrics, Neurology, and Neuroscience, University of Florida
carnepr@peds.ufl.edu

*This work was supported by the Department of Veterans Affairs and grant from NIH/NIBIB (8R01EB002089-03).

Abstract Epilepsy is a common neurological disorder characterized by recurrent seizures, most of which appear to occur spontaneously as a result of complex dynamical interactions among many regions of the brain. The most common type of epilepsy in adults is temporal lobe epilepsy. Employment of nonlinear dynamics techniques, based on the chaos theory, led us to the hypotheses that (1) seizures are a transition from spatiotemporal chaos to a more ordered ictal (symptomatic) state, (2) during the interictal (asymptomatic) state, the epileptogenic focus is dynamically isolated from other areas of the cerebral hemispheres, (3) the seizure discharge can occur only after a preictal transition state during which the focus starts to interact with other cortical areas (preictal state), and (4) the seizure serves to reset the brain, reversing the pathological interaction among critical cortical sites, thus dynamically isolating the seizure onset zone.

Through the analysis of long-term intracranial EEG recordings obtained in patients with medically intractable seizures, we observed: (1) drop in values of a dynamical measure (STL_{max}) during the seizure, indicating increased temporal order of the EEG signal – hypothesis 1, (2) convergence of STL_{max} values among almost all electrode pairs, indicating increased spatial order – hypothesis 1, (3) generally higher values of T-index (divergent values of STL_{max}) during the interictal state when the epileptogenic focus is compared to other sites – hypothesis 2, (4) the average differences in STL_{max} values between the epileptogenic hippocampus and most other sites is typically reduced (low T-index values, "dynamical entrainment") prior to each seizure – hypothesis 3, and (5) a divergence of STL_{max} (increase in the T-index) between the epileptogenic hippocampus and other cortical sites after each seizure – hypothesis 4. These observations support our hypotheses regarding the preictal transitions in temporal lobe epilepsy. We anticipate that these observations will lead to a better understanding of the physiological processes involved in temporal lobe epilepsy.

Keywords: Temporal Lobe Epilepsy, Short-Term Maximum Lyapunov exponents, T-Index, Entrainment Transition

1. Introduction

Epilepsy is one of the most common neurological disorders in man. Temporal lobe epilepsy is the most common type of epilepsy in adults [11]. In this disorder, seizures usually begin as paroxysmal electrical discharges in the hippocampus. The discharges often spread first to ipsilateral, then to contralateral cerebral cortex. These abnormal discharges result in a variety of intermittent clinical phenomena, including motor, sensory, affective, cognitive, autonomic and psychic symptomatology. This type of epilepsy often is resistant to medical therapy. In medically refractory cases, surgical excision of the seizure focus may be effective means of seizure control. In some surgical candidates, electrographic recordings are obtained for diagnostic purposes from subdural electrodes placed over the frontal and temporal cortex and depth electrodes implanted in the hippocampi bilaterally [29]. Such recordings offer a unique opportunity for research into electrophysiological processes of epileptogenesis in man.

In human epilepsy of mesial temporal origin, seizures beginning in the hippocampus are often propagated throughout the brain. The temporal cortex, limbic structures and orbitofrontal cortex appear to play a critical role in the onset and spread of these seizures [29]. Most medically intractable complex partial seizures originate from, or are elaborated in the hippocampus [8,35,36]. The cellular mechanisms underlying seizures of hippocampal origin in humans are not completely understood. However, it has been established that there are characteristic patterns of neuronal loss and dendritic damage associated with alterations in neurotransmitter receptor densities in the epileptogenic hippocampus [2,3,26,28,34]. It is likely that these structural changes disrupt the normal excitatory and inhibitory feedback circuits in the hippocampus, leading to disturbances in its dynamical behavior. A central feature of the epileptogenic hippocampus is the tendency to make abrupt transitions to well organized oscillations, characteristic of a seizure.

Successful surgical treatment of seizures of focal origin depends upon accurate presurgical localization of the epileptogenic focus. However, in some cases, accurate localization is not possible with noninvasive techniques. In such cases, electrographic recordings with surgically implanted subdural and/or depth electrodes are employed to obtain accurate localization. This procedure typically requires several days to weeks of hospital stay during which anticonvulsant drugs are withdrawn in order to allow seizures to occur. Typically, it is necessary to record 3 or more representative seizures in order to identify and localize the seizure focus. This time consuming and expensive process could be shortened considerably if the epileptogenic focus could be identified through analysis of the interictal electrographic signal.

The benefit of defining an epileptogenic focus through analysis of the interictal electrographic signal is clear. For every recorded seizure, there are many hours of interictal signal. The use of the interictal signal to localize epileptogenic foci requires identifying specific characteristics of the interictal signal generated by the epileptogenic focus. This objective has not been possible through visual inspection of the EEG. However, by detecting specific quantitative characteristics of the electrographic signal generated by the epileptogenic focus during the interictal state, it may be possible to identify and localize seizure foci. Our work to date strongly supports the view that the interictal signal generated by an epileptogenic focus has specific features that can be detected and quantified by analytic measures developed for the study of complex nonlinear systems.

Over the past several years, our group has sought to identify characteristic dynamical features of the electrographic signals generated by the epileptogenic focus. The possibility that such dynamical features might exist was suggested by the fact that there are structural abnormalities in the epileptogenic hippocampus (e.g., neuronal loss, dendritic abnormalities, and axonal sprouting of dentate granule cells) [3, 7, 10, 12, 27, 28, 33, 34] and there are localized zones of hypometabolism detectable with PET scans which are present during the interictal state [1, 25, 32]. Using techniques developed to identify nonlinearities in the signal recorded from depth and subdural electrodes in patients with temporal lobe epilepsy, we found that signals generated by the epileptogenic focus were strongly nonlinear [4–6]. We also found nonlinearities in signals generated by interictal spike foci located in the temporal cortex contralateral to the epileptogenic focus.

Traditionally, the initial occurrence of characteristic focal rhythmic EEG discharge, most commonly in the hippocampus or mesial temporal cortex, is considered to be the onset of a seizure. However, we have discovered, through analysis of the spatiotemporal dynamics of invasive electrographic recordings in patients with medically intractable temporal lobe epilepsy, a preictal transition process [13–22, 24, 31]. The onset of this transition precedes the seizure for periods of up to 1 hour. This transition was remarkably similar for each of the 48 seizures analyzed (4 seizures in each of 12 patients) and was not observed in the interictal data examined. The preictal dynamical transition is characterized by the progressive convergence of the mean of the STL_{max} values among specific anatomical areas (mean value entrainment) at specific times. For each case analyzed, this gradual preictal entrainment process culminated in a seizure. The discovery of a preictal transition period that can be detected by its quantitative dynamical characteristics offers the possibility of predicting an impending seizure in time to intervene therapeutically in order to abort the transition and prevent a seizure from occurring. If seizures can be detected minutes in advance, it could lead to the development of novel therapeutic approaches designed to disrupt the preictal transition process such as by electrical stimulation (e.g. vagal nerve stimulation) or by timely release of an anticonvulsant drug.

In this chapter, we emphasize the use of STL_{max} as a measure of the dynamical state of the electrographic recordings to investigate the interactions among brain areas during the interictal, preictal, ictal and postictal states. The Lyapunov exponent is a measure of the rate of production or destruction of information (usually expressed in bits per second) and is an indicator of how ordered (negative Lyapunov exponent) or chaotic (positive Lyapunov exponent) the series or the system is over a given time period. There are well established techniques for estimating the value of L_{max} in a time series. The EEG signal can be considered as a time series generated by a multi-dimensional system, the brain. In a multi-dimensional signal, as many Lyapunov exponents can be defined as there are dimensions [9]. A chaotic signal, by definition, has at least one positive Lyapunov exponent. The largest Lyapunov exponent (L_{max}) can be used to partially characterize the dynamical steady state of a physical system. In this application, the maximum Lyapunov exponent was estimated sequentially in each non-overlapping 10.24 sec EEG recording for each recorded cortical site. The interactions among brain areas were then investigated by generating the pair-T statistic time series of the maximum Lyapunov exponent values. We will utilize these observations to test the hypotheses: (1) seizures are a transition from spatiotemporal chaos to a more ordered ictal (symptomatic) state, (2) during the interictal (asymptomatic) state, the epileptogenic focus is dynamically isolated from other areas of the cerebral hemispheres, (3) the seizure discharge can occur only after a preictal transition state during which the focus starts to interact with other cortical areas (preictal state), and (4) the seizure serves to reset the brain, reversing the pathological interaction among critical cortical sites, thus dynamically isolating the seizure onset zone.

The rest of the chapter is organized as follows: In section 2, the algorithms for estimating STL_{max} and pair-T statistic are described. In section 3, the analyzed EEG recordings and the results of the analysis are presented. Section 4 gives the discussion of the results.

2. Nonlinear Dynamical and Statistical Measures

2.1. Nonlinear Dynamical Measure:

Iasemidis and Sackellares [22] applied the method of delays developed by Packard et al. and Takens [30,37] to reconstruct a multidimensional state space from a single-channel EEG signal. After the reconstruction by an embedding, each state is represented in the state space by a vector Y_t whose components are the delayed versions of the original single-channel EEG time series u_t, that is:

$$Y_t = [u_t, u_{t-\tau}, \ldots, u_{t-(p-1)\tau}] \quad (1)$$

where Y_t is a vector in the state space at time t, τ is the time delay between successive components of Y_t, and p is the embedding dimension of the reconstructed state space. The embedding dimension p is the dimension of the state space that contains the steady state of the system (i.e. attractor) and it is always a positive integer. On the other hand, the attractors' dimension D may be a positive non-integer (fractal). D is directly related to the number of variables of the system and is usually inversely related to the existing coupling among them. According to Takens [37], the embedding dimension p should be at least equal to $(2D + 1)$ in order to correctly embed an attractor in the state space. Each of many different methods used to estimate D of an object in the state space has its own practical problems. The measure most often used to estimate D is the state space correlation dimension ν. Methods for calculating ν from experimental data have been described and were employed in our work to approximate D of the epileptic attractor. In the EEG data we have analyzed to date, ν is found to be between 2 and 3 during an epileptic seizure. Therefore, in order to capture characteristics of the epileptic attractor, we have used an embedding dimension p of 7 for the reconstruction of the state space.

Since the brain is a nonstationary system, algorithms used to estimate measures of the brain dynamics should be capable of automatically identifying and appropriately weighing existing transients in the data. The method we developed for estimation of L_{max} for nonstationary data, called STL (Short-Term Lyapunov), considers possible nonstationarities in the EEG. This method was expained in details in Iasemidis et al. [22] We apply the STL algorithm to EEG tracings from electrodes in multiple brain sites, to create a set of STL_{max} time series. This set of time series contains local (in time and in space) information about the brain as a dynamical system. It has been shown that it is at this level of spatiotemporal analysis that reliable detection of the transition to epileptic seizures, long before they actually occur, is derived [15,23,24,31].

2.2. Statistical Determination of Dynamical Entrainment among Brain Areas:

Dynamical entrainment is defined as the convergence of STL_{max} values among the electrode sites within a window. This convergence is quantified by the average of pair-T statistics over all pairs among the group of sites. We defined this value as average T-index. The calculation of a pair-T statistic is described as follows:

For electrode sites i and j, let their STL_{max} values in a window W_t of n STL_{max} points are

$$L_i^t = \{STL_{max_i}^t, STL_{max_i}^{t+1}, \ldots, STL_{max_i}^{t+n-1}\} \tag{2}$$

$$L_j^t = \{STL_{max_j}^t, STL_{max_j}^{t+1}, \ldots, STL_{max_j}^{t+n-1}\} \tag{3}$$

and

$$D_{ij}^t = L_i^t - L_j^t = \{d_{ij}^t, d_{ij}^{t+1}, \ldots, d_{ij}^{t+n-1}\} \tag{4}$$

Then, the pair-T statistic at time window W_t between electrode sites i and j is calculated by

$$T_{ij}^t = \frac{|\bar{D}_{ij}^t|\sqrt{n}}{\hat{\sigma}_d} \tag{5}$$

where $\bar{D}_{ij}^t$ and $\hat{\sigma}_d$ are the sample mean value and the sample standard deviation of D_{ij}^t.

If the true mean of $\bar{D}_{ij}^t$, denoted by μ_{ij}^t, is equal to zero, and the assumptions of $\bar{D}_{ij}^t$ being independent and normal-distributed are valid [23], asymptotically, T_{ij}^t is distributed as a t-distribution with $n-1$ degrees of freedom.

If we define the disentrainment between electrode sites i and j as μ_{ij}^t is significantly different from zero with significance level α, the disentrainment between two electrode sites can then be detected by the pair-T test as: Electrode sties i and j are disentrained if $T_{ij}^t > t_{\alpha/2,n-1}$, where $t_{\alpha/2,n-1}$ is the $100(1-\alpha/2)\%$ critical value of t-distribution with $n-1$ degrees of freedom. If $T_{ij}^t < t_{\alpha/2,n-1}$, which means that the differences of STL_{max} values between electrode sites i and j in the time window W_t do not have sufficient evidence to claim the disentrainment between these two sites, in this situation, we will consider sites i and j are entrained each other in W_t.

T-index profiles were generated over time for all possible pairs. Average T-index curve between two brain areas was created by averaging all the possible pairs between these two brain areas. For example, the average T-index curve between LTD (left temporal depth) and RTD (right temporal depth) over time is the average of all possible pairs (6 LTD's $\times$ 6 RTD's = 36 pairs) between these two areas. When the average T-index is small, it indicates that the two brain areas are more dynamically interacted (entrained) to each other at that particular time window with respect to their STL_{max} values.

3. Results

Electrographic recordings from bilaterally, surgically implanted microelectrodes in the hippocampus, temporal and frontal lobe cortexes of an epileptic patient with temporal lobe, complex, focal, with secondarily generalized epileptic seizures, was analyzed (see Figure 12.1 for our typical electrode montage). The EEG signals were recorded using amplifiers with an input range of $\pm 0.6mV$, and a frequency range of $0.5 \sim 70$ Hz. Prior to storage, the signals were sampled at 200 Hz using an analog to digital (A/D) converter with 10-bit quantization. The multi-electrode EEG signals (28 common reference channels) were obtained from long-term (11.67 hours) continuous recordings from this patient. Five seizures of mesial temporal onset were recorded during the period of recordings. This EEG recording was viewed by two independent electroencephalographers to determine the number and type of recorded seizures, seizure onset and end times, and seizure onset zones. The area RTD (right temporal depth) was identified as the epileptogenic focal area of this patient.

Figure 12.2 demonstrates a typical STL_{max} profile from an epileptogenic focal cortical site over 3 hours including two seizures and 1 hour after the second seizure. The STL_{max} values remain high during the interictal period (more chaotic), gradually start decreasing approximately 30 minutes before the seizure (more ordered), and drop to the lowest points during the ictal periods (most ordered). These observations support the hypothesis that seizures are a transition from temporal chaos to a more ordered ictal (symptomatic) state.

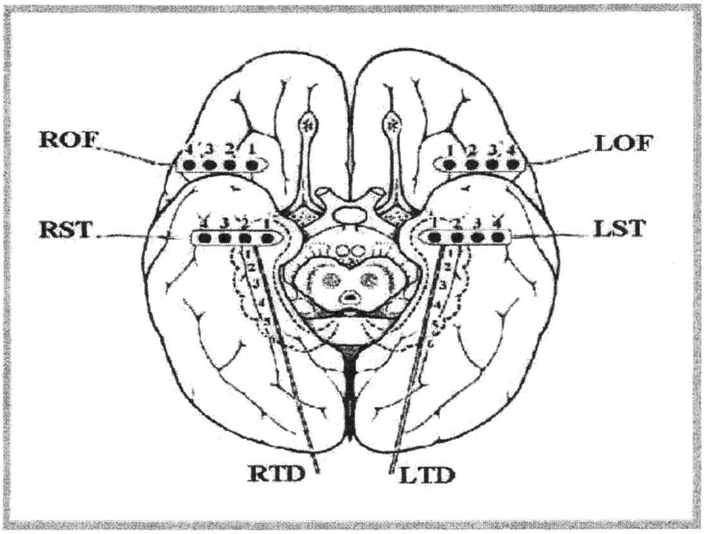

Figure 12.1. Graphic illustration of placement of subdural electrode strips and depth electrodes used for long-term diagnostic intracranial EEG recordings. Electrode strips are placed over the left orbitofrontal (LOF), right orbitofrontal (ROF), and left subtemporal (LST) and right subtemporal (RST) cortex. Depth electrodes are placed in the left temporal (LTD) and right temporal (RTD) lobes to record electrical activity generated by the amygdala and hippocampus.

Figure 12.3 shows the percentage of entrained electrode pairs matched from different brain areas with significance level $\alpha = 0.05$. Four electrodes from each area were included in the analysis. Therefore, there are total 240 electrode pairs. The figure showed that during the ictal periods, the percentage of entrained electrode pairs significantly increases to the maximum values, and drop to the lowest points approximately 10 minutes after the seizure. The observation of massive convergence of STL_{max} (entrainment) during the ictal period indicates increased spatial order and thus supports our hypothesis 1. Further, the observation that less entrainment occurs after the seizure indicates that seizure serves to reset the brain, reversing the pathological interaction among critical cortical sites (hypothesis 4).

Figures 12.4 and 12.5 show the T-index profiles between the epileptogenic focus (RTD/RST) and other areas (LTD, LST, LOF and ROF) of the cerebral hemispheres. During the interictal state, generally higher values of T-index (divergent values of STL_{max}) indicate that the epileptogenic focus is dynamically isolated from other areas of the cerebral hemispheres. This observation supports our hypothesis 2. Further, the average differences in STL_{max} values between the epileptogenic hippocampus and most other sites are typically

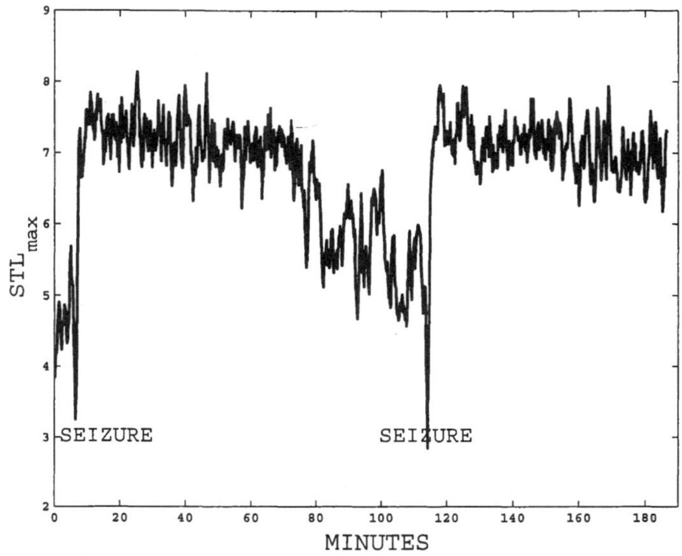

Figure 12.2. STL_{max} profiles over 3 hours including two seizures and 1 hour after the second seizure. Using embedding dimension p=7 and time delay τ=20 msec for the state space reconstruction, the STL_{max} values were estimated by dividing the EEG signal into non-overlapping epochs of 10.24 seconds each.

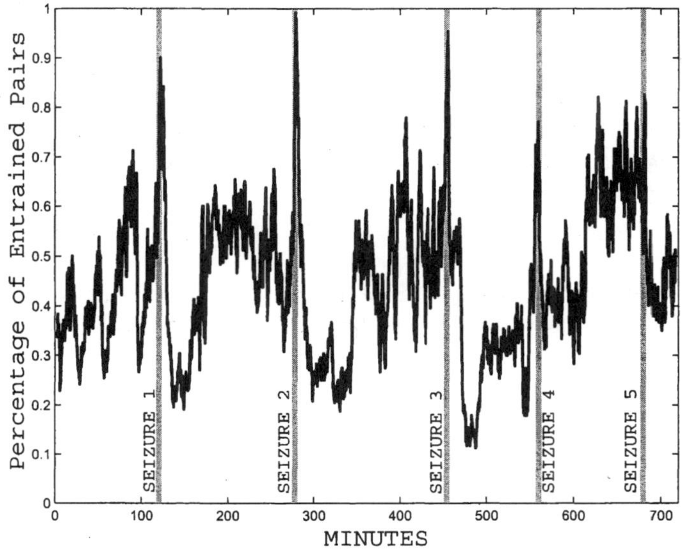

Figure 12.3. Percentage of entrained ($\alpha = 0.05$) electrode pairs matched in different brain areas.

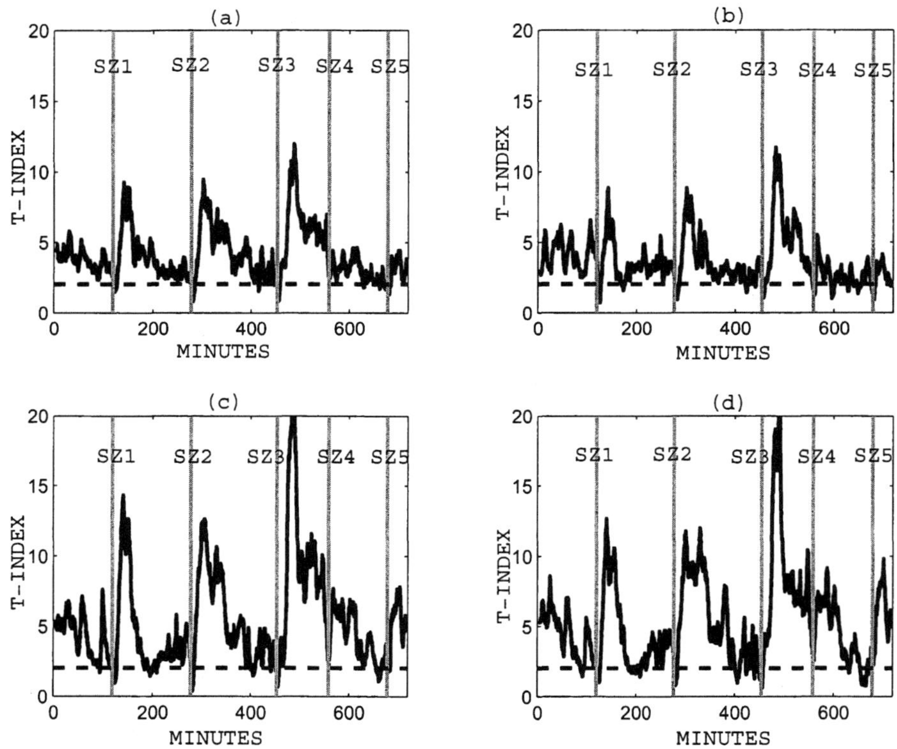

Figure 12.4. T-index profiles between RTD and other areas of the cerebral hemispheres over 12 hours including 5 seizures: (a) RTD versus LTD, (b) RTD versus LST, (c) RTD versus LOF, and (d) RTD versus ROF.

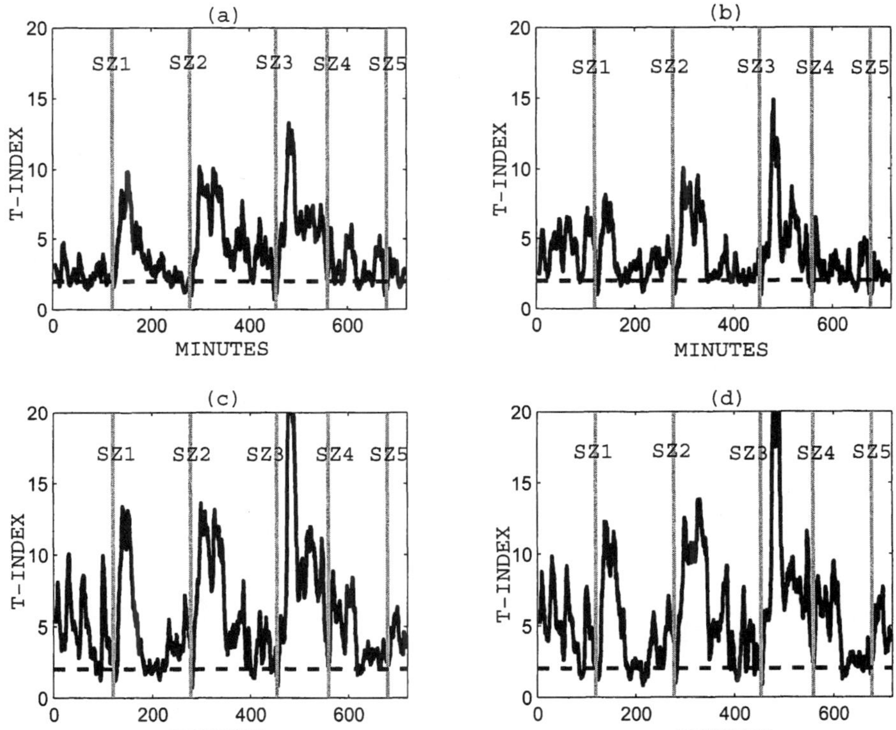

Figure 12.5. T-index profiles between RST and other areas of the cerebral hemispheres over 12 hours including 5 seizures: (a) RST versus LTD, (b) RST versus LST, (c) RST versus LOF, and (d) RST versus ROF.

reduced (low T-index values, "dynamical entrainment") prior to each seizure, which supports the hypothesis 3. The fourth hypothesis is also supported by a divergence of STL_{max} (increase in the T-index) between the epileptogenic hippocampus and other cortical sites after each seizure. In other words, the dynamical entrainment between the epileptogenic focus and other areas is reset (disentrained) by the occurrences of the seizures.

4. Discussion

In this chapter, we hypothesize the state transitions of the temporal lobe epilepsy during the interictal, preictal, ictal and postictal states. These hypotheses are supported by the observations based on the dynamical and statistical analysis of the EEG signals. These studies suggest that a seizure is a spatiotemporal transition and occurs only when a sufficiently large area of cerebrum becomes entrained dynamically with the epileptogenic focus. Although these observations could lead us toward the understanding of the transitions in temporal lobe epilepsy, however, the questions such as how the dynamical entrainment occurs and how it develops prior to a seizure are still remained unanswered. Fully understanding of these mechanisms is the key to develop a reliable epileptic seizure predictor and to control or prevent an impending seizure well before its actual onset.

References

[1] B.W. Abou-Khalil, G.J. Siegel, J.C. Sackellares, S. Gilman, R. Hichwa, and R. Marshall. Positron emission tomography studies of cerebral glucose metabolism in patients with chronic partial epilepsy. *Ann. Neurol.*, 22:480–486, 1987.

[2] T.L. Babb and W.J. Brown. Pathological findings in epilepsy. In J. Engel Jr., editor, *Surgical treatment of the epilepsies*. Raven Press, 1987.

[3] D.E. Burdette, Sakuraisy, T.R. Henry, D.A. Ross, P.B. Pennell, J.C. Sackellares K.A. Frey, and R. Albin. Temporal lobe central benzodiazepine binding in unilateral mesial temporal lobe epilepsy. *Neurology*, 45:934–941, 1995.

[4] M.C. Casdagli, L.D. Iasemidis, R.L. Gilmore, S.N. Roper, R.S. Savit, and J.C. Sackellares. Nonlinearity in invasive eeg recordings from temporal lobe epilepsy. *EEG Clin. Neurophysiology*, 102:98–105, 1997.

[5] M.C. Casdagli, L.D. Iasemidis, J.C. Sackellares, S.N. Roper, R.L. Gilmore, and R.S. Savit. Characterizing nonlinearity in invasive eeg recordings from temporal lobe epilepsy. *Physica D*, 99:381–399, 1996.

[6] M.C. Casdagli, L.D. Iasemidis, R.S. Savit, R.L. Gilmore, S.N. Roper, and J.C. Sackellares. Nonlinear analysis of mesial temporal lobe seizures using a surrogate data technique. *Epilepsia*, 36:142, 1995.

[7] J.A.N. Corsellis and B.S. Meldrum. Epilepsy. In W. Blackwood and J.A.N. Corsellis, editors, *Neuropathology*, pages 771–795. Arnold, 1976.

[8] V.M. Fernandes de Lima, J.P. Pijn, C.N. Filipe, and F. Lopes da Silva. The role of hippocampal commissures in the interhemispheric transfer of epileptiform afterdischarges in the rat: a study using linear and nonlinear regression analysis. *Electroenceph. Clin. Neurophysiol.*, 76:520–539, 1990.

[9] J.P. Eckmann, S.O. Kamphorst, D. Ruelle, and S. Ciliberto. Lyapunov exponents from time series. *Phys. Rev. A*, 34:4971–4972, 1986.

[10] M.A. Falconer, E.A. Serefetinides, and J.A.N. Corsellis. Aetiology and pathogenesis of temporal lobe epilepsy. *Arch. Neurol.*, 19:233–240, 1964.

[11] H. Gastaut, J.L. Gastaut, G.E. Concalves de Silva, and G.R. Fernandez Sanchez. Relative frequency of different types of epilepsy: a study employing the classification of the international league against epilepsy. *Epilepsia*, 16(3):457–461, 1975.

[12] T.R. Henry, K.A. Frey, J.C. Sackellares, S. Gilman, R.A. Koeppe, J.A. Brunberg, D.A. Ross, S. Berent, H.A. Buchtel, A.B. Young, and D.E. Kuhl. In vivo cerebral metabolism and central benzodiazapine receptor binding in temporal lobe epilepsy. *Neurology*, 43:1998–2006, 1993.

[13] L.D. Iasemidis, A. Barretov R.L. Gilmore, B.M. Uthman, S. Roper, and J.C. Sackellares. Spatiotemporal evolution of dynamical measures precedes onset of mesial temporal lobe seizures. *Epilepsia*, 35(suppl. 8):133, 1994.

[14] L.D. Iasemidis, L.D. Olson, R.S. Savit, and J.C. Sackellares. Time dependencies in the occurrences of epileptic seizures. *Epilepsy Research*, 17:81–94, 1994.

[15] L.D. Iasemidis, P.M. Pardalos, J.C. Sackellares, , and D.-S. Shiau. Quadratic binary programming and dynamical system approach to determine the predictability of epileptic seizures. *Journal of Combinatorial Optimization*, 5:9–26, 2001.

[16] L.D. Iasemidis, J.C. Principe, J.M. Czaplewski, R.L. Gilman, S.N. Roper, and J.C. Sackellares. Spatiotemporal transition to epileptic seizures: A nonlinear dynamical analysis of scalp and intracranial eeg recordings. In J.C. Principe F. L.Silva and L.B. Almeida, editors, *Spatiotemporal Models in Biological and Artificial Systems*. IOS Press, 1997.

[17] L.D. Iasemidis and J.C. Sackellares. Long time scale temporo-spatial patterns of entrainment of preictal electrocorticographic data in human temporal lobe epilepsy. *Epilepsia*, 31(5):621, 1990.

[18] L.D. Iasemidis and J.C. Sackellares. The evolution with time of the spatial distribution of the largest lyapunov exponent on the human epileptic cortex. In D.W. Duke and W.S. Pritchard, editors, *Measuring Chaos in the Human Brain*. World Scientific, 1991.

[19] L.D. Iasemidis and J.C. Sackellares. The use of dynamical analysis of eeg frequency content in seizure prediction. *EEG Clin. Neurophysiology*, 1993.

[20] L.D. Iasemidis and J.C. Sackellares. Chaos theory and epilepsy. *The Neuroscientist*, 12:118–126, 1996.

[21] L.D. Iasemidis, J.C. Sackellares, and R.S. Savit. Quantification of hidden time dependencies in the eeg within the framework of nonlinear dynamics. In B. Jansen, editor, *Nonlinear Dynamical Analysis of the EEG*. World Scientific Publishing Company, 1993.

[22] L.D. Iasemidis, J.C. Sackellares, H.P. Zaveri, and W.J. Williams. Phase space topography of the electrocorticogram and the lyapunov exponent in partial seizures. *Brain Topogr.*, 2:187–201, 1990.

[23] L.D. Iasemidis, D.-S. Shiau, W. Chaovalitwongse, J.C. Sackellares, P.M. Pardalos, J.C. Principe, P.R. Carney, A. Prasad, B. Veeramani, and K. Tsakalis. Adaptive epileptic seizure prediction system. *IEEE Transactions on Biomedical ENgineering*, 50(5):616–627, 2003.

[24] L.D. Iasemidis, D.-S. Shiau, J.C. Sackellares, and P.M. Pardalos. Transition to epileptic seizures: Optimization. In D.Z. Du, P.M. Pardalos, and J. Wang, editors, *DIMACS series in Discrete Mathematics and Theoretical Computer Science*. American Mathematical Society, 1999.

[25] J. Engel Jr., D.E. Kuhl, M.E. Phelps, and J.C. Mazziota. Interictal cerebral glucose metabolism in partial epilepsy and its relation to eeg changes. *Ann. Neurol.*, 12:510–517, 1982.

[26] N.C. De Lanerrole, J.H. Kim, R.J. Robbins, and D.D. Spencer. Hippocampal interneuron loss and plasticity in human temporal lobe epilepsy. *Brain. Res.*, 495:387–395, 1989.

[27] J.H. Margerison and J.A.N. Corsellis. Epilepsy and the temporal lobes. *Brain*, 89:499–530, 1966.

[28] J.W. McDonald, E.A. Garofalo, T. Hood, J.C. Sackellares, P.E. McKeever S. Gilman, J.C. Troncaso, and M.V. Johnston. Altered excitatory and inhibitory aminoacid receptor binding in hippocampus of patients with temporal lobe epilepsy. *Annals of Neurology*, 29:529–541, 1991.

[29] E. Niedermeyer. Depth electroencephalography. In E. Niedermeyer and F. Lopes da Silva, editors, *Electroencephalography: Basic principles, clinical applications and related fields*, pages 593–617. Urban and Schwarzenberg, 1987.

[30] N.H. Packard, J.P. Crutchfield, J.D. Farmer, and R.S. Shaw. Geometry from time series. *Phys. Rev. Lett.*, 45:712–716, 1980.

[31] J.C. Sackellares, L.D. Iasemidis, D.-S. Shiau, R.L. Gilmore, and S.N. Roper. Epilepsy - when chaos fails. In K. Lehnertz, J. Arnhold, P. Grassberger, and C.E. Elger, editors, *Chaos in the brain?* World Scientific, 2000.

[32] J.C. Sackellares, G.J. Siegel, B.W. Abou-Khalil, T.W. Hood, P. McKeever S. Gilman, R.D. Hichwa, and G.D. Hutchins. Differences between lateral and mesial temporal metabolism interictally in epilepsy of mesial temporal origin. *Neurology*, 40:1420–1426, 1990.

[33] H.J. Sagar and J.M. Oxbury. Hippocampal neuron loss in temporal lobe epilepsy: correlation with early childhood convulsions. *Ann Neurol.*, 22:334–340, 1987.

[34] I. Savic, A. Persson, P. Roland, S. Pauli, G. Dedvall, and L. Witen. In vivo demonstration of reduced benzodiazepine receptor binding in human epileptic foci. *Lancet*, 2:863–866, 1988.

[35] D.D. Spencer and C.T.E. Pappas. Surgical decisions regarding medically intractable epilepsy. *Clinical Neurosurgery*, 38:548–566, 1992.

[36] S.S. Spencer, K. Jung, and D.D. Spencer. Ictal spikes: a marker of specific hippocampal cell loss. *Electroenceph. Clin. Neurophysiol.*, 83:104–111, 1992.

[37] F. Takens. Detecting strange attractors in turbulence. In D.A. Rand and L.S. Young, editors, *Dynamical systems and turbulence, Lecture Notes in Mathematics*, pages 366–381. Springer-Verlag, 1981.

Chapter 13

NONLINEAR DYNAMICAL AND STATISTICAL APPROACHES TO INVESTIGATE STATE TRANSITIONS BEFORE EPILEPTIC SEIZURES *

D.-S. Shiau
Department of Neuroscience, University of Florida; VA Medical Center, Gainesville, FL
shiau@epilepsy.health.ufl.edu

W. Chaovalitwongse
Department of Industrial and Systems Engineering, University of Florida; VA Medical Center, Gainesville, FL
arty@epilepsy.health.ufl.edu

L.D. Iasemidis
Department of Biomedical Engineering, Arizona State University
Leon.Iasemidis@asu.edu

P.M. Pardalos
Departments of Industrial and Systems Engineering, Computer Science, and Biomedical Engineering, University of Florida
pardalos@cao.ise.ufl.edu

P.R. Carney
Department of Pediatrics, Neurology, and Neuroscience, University of Florida
carnepr@peds.ufl.edu

*This work was supported by the Department of Veterans Affairs and grant from NIH/NIBIB (8R01EB002089-03).

J.C. Sackellares
Departments of Neuroscience, Neurology, Pediatrics, and Biomedical Engineering, University of Florida; VA Medical Center, Gainesville, FL
sackellares@mbi.ufl.edu

Abstract Epilepsy is a dynamical disorder of the brain. Initial results obtained, from a small sample of patients, by employing signal processing techniques, based on the theory of nonlinear dynamics and statistics, led us to hypothesize that the dynamical entrainment (i.e., their dynamical measures are gradually convergent in time) among the critical cortical brain sites can be used to anticipate an impending seizure [10, 17]. In this chapter, we present the results from a larger sample of patients and seizures to confirm this hypothesis.

Through the analysis of long-term intracranial EEG recordings obtained in 10 patients with 118 medically intractable seizures, we observed that seizures were preceded by a preictal transition that evolves over approximately 54 minutes. This transition is followed by a seizure. The study of this process has been hampered by its complexity and variability. A major problem is that the transitions involve a subset of brain sites that vary from seizure to seizure, even in the same patient, which is expected since the state of the patient may be different before each seizure. However, by combining dynamical analytic techniques with a critical cortical site selection method, we have been able to elucidate important dynamical characteristics underlying human epilepsy. We anticipate that these observations will lead to a better understanding of the physiological processes involved. We illustrate the use of these approaches in confirming our hypotheses regarding entrainment characteristics prior the seizure. Thus, it may be possible to develop novel therapeutic approaches involving carefully timed interventions to prevent the patients from the occurrence of a seizure.

Keywords: Epilepsy, nonlinear dynamics, Lyapunov exponents, T-index, Preictal Transition, Seizure predictability

1. Introduction

Epilepsy is the most common serious brain disorder in every country of the world. It may be the most universal of all medical disorders, affecting all ages, races, social classes, and nations [18]. Worldwide, at least 40 million or 7 of every 1,000 individuals currently suffer from epilepsy [2]. Estimates of incidence rates range from 24 to 53 per 100,000 [1, 3, 8, 9, 19, 21]. The mainstay of contemporary treatment for epilepsy is pharmacological. Anticonvulsant drugs are taken daily, in fixed doses, and are titrated to achieve a steady-state concentration in the blood. One problem with chronic daily dosing with anticonvulsant drugs is that many patients develop a tolerance to the anticonvulsant effect. This is particularly true for the most powerful class of anticonvulsants, the benzodiazepines. During the past decade several new anticonvulsants have been available for adults and children with epilepsy. Nonetheless, approximately 25% of these individuals have seizures that are refractory to medical therapy [5]. For these patients, surgical treatment may be an option. Surgical treatment can be effective in carefully selected cases. Good responses occur in approximately 70 to 90% of adults with temporal lobe epilepsy [5]. However, therapeutic response rates drop off markedly in individuals with more than one epileptogenic focus or those with generalized seizures.

Among the most disabling aspects of epilepsy are the anticipation of seizures and the uncertainty of when the next seizure will occur. Patients with epilepsy usually appear normal and function normally for much of the time. Then suddenly a seizure occurs. The central questions as to why seizures occur intermittently, and when they begin and end, remain unanswered.

Our previous studies have shown that seizures are not abrupt events; instead, they follow a dynamical transition that evolves over minutes to hours before the seizures [10–17, 22]. During this preictal dynamical transition, multiple regions of the cerebral cortex progressively approach a similar dynamical state. Other investigations have confirmed the presence of a progressive preictal transition [4,6,7,20]. For seizures of frontal and temporal lobe origin, the dynamical transition involves the gradual convergence of the dynamical characteristics of multiple areas of the cerebral cortex toward a similar state. This preictal convergence may involve different brain regions from seizure to seizure even in the same patient. Therefore, accurate detection of the preictal transition depends upon selecting the appropriate cortical sites. In this chapter, we present a preictal transition detection algorithm by combining the nonlinear dynamics and statistical analyses with a critical cortical sites selection method.

The rest of the chapter is organized as follows: In section 2, the measures and the methodology employed for testing the seizure predictability hypotheses are described. In section 3, the EEG data and the results of the data analysis are presented. Discussion of the results is provided in final section 4.

2. Measures and Methods

2.1. Dynamical Measure

We utilized an estimate of the Short-Term Maximum Lyapunov exponent (STL_{max}) as the dynamical measure of the electroencephalogram. Estimation of STL_{max} was calculated by dividing the EEG signal into non-overlapping segments of 10.24-sec each. The largest Lyapunov exponent (L_{max} or L_1) is defined as the average of local Lyapunov exponents L_{ij} in the state space, that is:

$$L_{max} = \frac{1}{N_\alpha} \sum_{N_\alpha} L_{ij} \tag{1}$$

where N_α is the total number of the local Lyapunov exponents that are estimated from the evolution of adjacent points (vectors) in the state space. $X_i = X(t_i), X_j = X(t_j)$, and

$$L_{ij} = \frac{1}{\Delta t} \times \log_2 \frac{|X(t_i + \Delta t) - X(t_j + \Delta t)|}{|X(t_i) - X(t_j)|} \tag{2}$$

where Δt is the evolution time allowed for the vector difference $|X(t_i) - X(t_j)|$ to evolve to the new difference $|X(t_i+\Delta t) - X(t_j+\Delta t)|$. If Δt is given in sec, then STL_{max} should be in bits/sec. More details regarding STL_{max} can be found in Iasemidis et al., 1990 [16]. Figure 13.1 shows a 20-minute EEG recording, 10 minutes before and after a seizure onset, and its corresponding STL_{max} curve.

2.2. Statistical T-index

We used the T-index (from the statistical paired-T test) to measure the degree of entrainment (convergence with respect to the STL_{max} values) between cortical sites. The T-index of a pair of sites was calculated in each 10-min sliding window (60 STL_{max} segments) by dividing the mean difference of STL_{max} values between the two cortical sites by its standard deviation. That is, the T-index at time t between cortical sites i and j is defined as:

$$T_{i,j}(t) = \sqrt{60} \times |E\{STL_{max,i} - STL_{max,j}\}| / \sigma_{i,j}(t) \tag{3}$$

where $E\{STL_{max,i}(t) - STL_{max,j}(t)\}$ denotes the average difference of STL_{max} between electrode sites i and j, within a 10-min time window, and $\sigma_{i,j}(t)$ is the sample standard deviation of the differences.

2.3. Critical Cortical Sites Selections

One of the most important tasks to detect the dynamical transitions is to identify the most possible group of cortical sites which will participate in the preictal transition of an impending seizure. Here we identify the most critical group of cortical sites based on the dynamical entrainment (T-index values) in the 10-min time window before the seizure. More specifically, the algorithm selects the group of cortical sites which are most entrained (minimum average T-indices) prior to the seizure. This task can be easily accomplished by creating a T-index matrix before the seizure. The objective in this chapter is to confirm the hypothesis that a preictal transition can be detected by observing the average T-index curve of the identified most critical group of cortical sites. Figure 13.2 show some examples of T-index curve from the identified most critical cortical sites before seizures. It is noticed that the T-index curves gradually decrease and drop below a critical value tens of minutes before seizures. These critical cortical sites remain entrained until the seizure occurs.

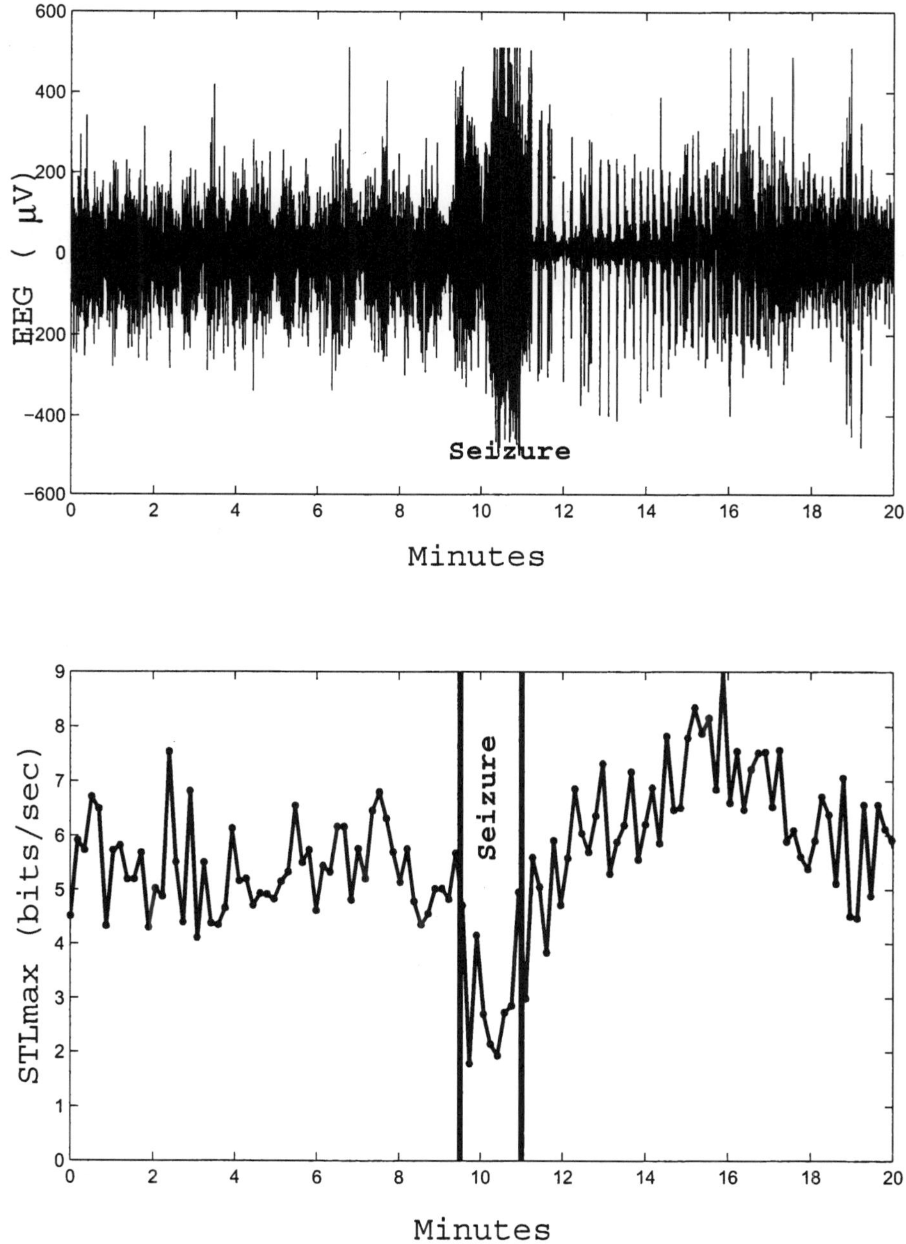

Figure 13.1. EEG and STL_{max} profiles over 20 minutes including a 1.5-minute ictal period. The estimation of the STL_{max} values was made by dividing the EEG signal into non-overlapping segments of 10.24 seconds each, using embedding dimension $p = 7$ and time delay $\tau = 20$ msec for the state space reconstruction.

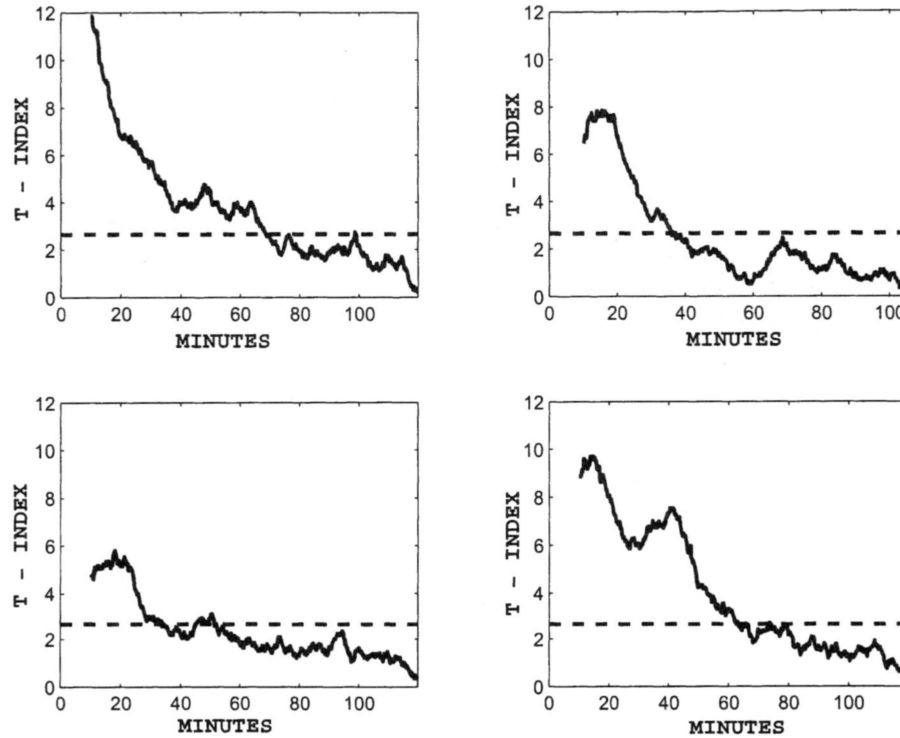

Figure 13.2. Four T-index curves from the identified groups of critical cortical sites before seizures.

2.4. Detection of Preictal Transitions

Based on the T-index curve described above, the decision of whether a preictal transition is detectable is made by the following steps:

(1) Sequentially averaging the T-index values, backwards in time from the seizure onset, to identify the first time point where the averaged T-index curve is above the critical value T_α (from the T-distribution for a given significance level α). Then, define the duration of the preictal transition (DPT_α) as the time interval from the previously identified time point to the seizure onset time point.

(2) After determining the DPT_α, the moving window length is set equal to DPT_α. The average T-index values within each overlapping moving window is then estimated backwards in time up to 2 hours prior to the seizure's onset. A false positive is observed if the average T-index within one of the windows is less than the critical value T_α that was used to determine the DPT_α.

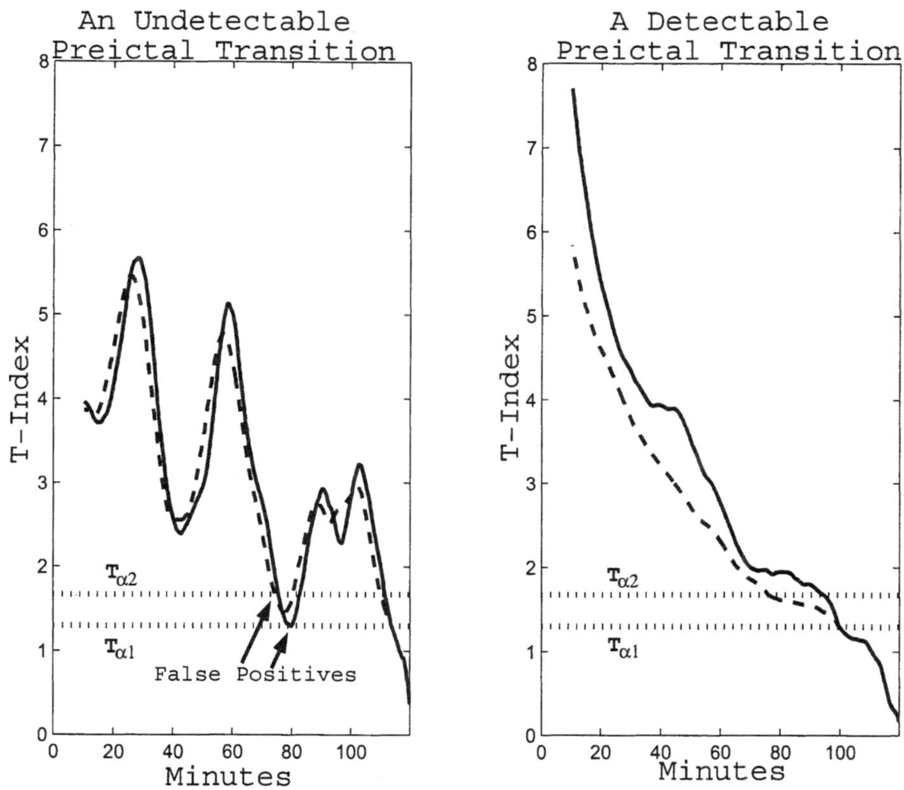

Figure 13.3. Illustrations of a detectable and an undetectable preictal transitions.

(3) Two critical values, α_1 and α_2, are used in this method. The hypothesis that the preictal transition can be detected is rejected if false positives are observed for both critical values either at the same or different time points in the available interval prior to the seizure. Otherwise, the preictal transition is considered detected. Figure 13.3 shows the illustrations of a detectable and an undetectable preictal transitions.

3. Results

In this section, results from the application of the previously described method to detect the preictal transitions of epileptic seizures are shown. The method was applied to 118 epileptic seizures in 10 patients. Table 13.1 summaries the statistics of the analyzed EEG data sets.

Table 13.2 gives the summary of the results of this analysis for all 118 seizures. The results show that more than 85% of the seizures have detectable

Table 13.1. Statistics of ten analyzed EEG data sets from ten patients. RTL = Right Temporal Lobe, LTL = Left Temporal Lobe.

Patient	Epileptic Focus	Total ♯ of Seizures	Mean Seizure Interval (Hrs)
♯1	RTL	24	3.62
♯2	RTL	19	7.77
♯3	LTL	8	3.23
♯4	LTL	4	2.17
♯5	RTL	3	4.15
♯6	RTL	7	20.32
♯7	RTL	9	8.69
♯8	RTL	17	17.10
♯9	RTL	18	15.30
♯10	RTL	9	7.59
Total		118	9.69

Table 13.2. Summary of the preictal transition detection analysis on 118 epileptic seizures in 10 patients.

Patient	Total ♯ of Seizure	Predictable Seizures	Detectability of Preictal Transitions (%)	Average Prediction Time (min.)
♯1	24	21	87.5	66.9
♯2	19	17	89.5	29.8
♯3	8	8	100.0	49.5
♯4	4	4	100.0	44.1
♯5	3	3	100.0	34.4
♯6	7	4	57.1	43.3
♯7	9	8	88.9	53.7
♯8	17	15	88.2	69.6
♯9	18	13	72.7	60.8
♯10	9	8	88.9	49.4
Total	118	101	85.6	53.9

preictal transitions. The average time interval of preictal transitions is approximately 54 minutes.

4. Discussions

This study was undertaken to confirm the hypothesis that it is possible to predict temporal lobe epileptic seizures by the analysis of dynamical and statistical characteristics of multi-channel EEG signals recorded from multiple cortical sites. Our previous studies from a small sample of patients have indicated the existence of preictal transitions starting 30 $\sim$ 60 minutes before a seizure onset, in which the values of Lyapunov exponents of EEG recorded from critical cortical sites become convergent. However, the cortical sites involved in this dynamical transition vary from seizure to seizure. Thus, the ability to identify the critical cortical sites that participate in the preictal transition plays a key role for the prediction of an impending seizure.

By employing a critical cortical site selection method which selects the most entrained group of cortical sites prior to a seizure, we demonstrated that 85.6% of the 118 seizures analyzed can be anticipated by the detection of their preictal transitions. These transitions can be observed on an average of approximately 54 minutes before the seizure onsets. These results, from a larger sample of patients and seizures, not only confirm our previous findings of the existence of the preictal transitions, but also give us more confidence to develop computer based automatic on-line, real time seizure prediction systems. Such systems not only can be used to enhance patient safety and treatment by alerting the nursing and technical staff of impending seizures, but also could provide promise for new diagnostic applications and novel approaches to seizure control.

References

[1] M. Brewis, D.C. Poskanzer, C. Rolland, and H. Miller. Neurological disease in an english city. *Acta Neurologica Scandinavica*, 42(24):9–86, 1966.

[2] O.C. Cockerell, I. Eckle, D.M. Goodridge, J.W. Sander, and S.D. Shorvon. Epilepsy in developing countries. *J. Neurol. Neurosurg. Psychiatry*, 58(5):570–576, 1995.

[3] A.S. de Graaf. Epidemiological aspects of epilepsy in northern norway. *Epilepsia*, 15(3):291–299, 1974.

[4] C.E. Elger and K. Lehnertz. Seizure prediction by non-linear time series analysis of brain electrical activity. *Europ. J. Neurosci.*, 10:786–789, 1998.

[5] J. Engel and T.A. Pedley. *Epilepsy: A comprehensive textbook*. Philadelph: Lippincott-Raven Pub, 1998.

[6] J. Martinerie et al. Epileptic seizures can be anticipated by non-linear analysis. *Nature Medicine*, 4:1173–1176, 1998.

[7] M.L.V. Quyen et al. Anticipation of epileptic seizures from standard eeg recordings. *Lancet*, 357:183–188, 2001.

[8] E. Granieri, G. Rosati, R. Tola, M. Pavoni, E. Paolino, L. Pinna, and V.C. Monetti. A descriptive study of epilepsy in the district of copparo, italy, 1964-1978. *Epilepsia*, 24:502–514, 1983.

[9] W.A. Hauser and L.T. Kurland. The epidemiology of epilepsy in rochester, minnesota, 1935 through 1967. *Epilepsia*, 16(1):1–66, 1975.

[10] L.D. Iasemidis, P.M. Pardalos, J.C. Sackellares, and D.-S. Shiau. Quadratic binary programming and dynamical system approach to determine the predictability of epileptic seizures. *Journal of Combinatorial Optimization*, 5:9–26, 2001.

[11] L.D. Iasemidis, J.C. Principe, J.M. Czaplewski, R.L. Gilman, S.N. Roper, and J.C. Sackellares. Spatiotemporal transition to epileptic seizures: A nonlinear dynamical analysis of scalp and intracranial eeg recordings. In F. L.Silva, J.C. Principe, and L.B. Almeida, editors, *Spatiotemporal Models in Biological and Artificial Systems*. IOS Press, 1997.

[12] L.D. Iasemidis, J.C. Principe, and J.C. Sackellares. Spatiotemporal dynamics of human epileptic seizures. In R.G. Harrison, L. Weiping, W. Ditto, L. Pecora, and S. Vohra, editors, *3rd Experimental Chaos Conference*. World Scientific, 1996.

[13] L.D. Iasemidis and J.C. Sackellares. Long time scale temporo-spatial patterns of entrainment of preictal electrocorticographic data in human temporal lobe epilepsy. *Epilepsia*, 31(5):621, 1990.

[14] L.D. Iasemidis and J.C. Sackellares. The evolution with time of the spatial distribution of the largest lyapunov exponent on the human epileptic cortex. In D.W. Duke and W.S. Pritchard, editors, *Measuring Chaos in the Human Brain*. World Scientific, 1991.

[15] L.D. Iasemidis and J.C. Sackellares. Chaos theory and epilepsy. *The Neuroscientist*, 12:118–126, 1996.

[16] L.D. Iasemidis, J.C. Sackellares, H.P. Zaveri, and W.J. Williams. Phase space topography of the electrocorticogram and the lyapunov exponent in partial seizures. *Brain Topogr.*, 2:187–201, 1990.

[17] L.D. Iasemidis, D.-S. Shiau, J.C. Sackellares, and P.M. Pardalos. Transition to epileptic seizures: Optimization. In D.Z. Du, P.M. Pardalos, and J. Wang, editors, *DIMACS series in Discrete Mathematics and Theoretical Computer Science*. American Mathematical Society, 1999.

[18] P. Jallon. Epilepsy in a population of 6000 re-examined: secular trends in first attendance rates, prevalence, and prognosis. *Epilepsia*, 38(10):1143–1151, 1997.

[19] P. Joensen. Prevalence, incidence, and classification of epilepsy in the faroes. *Acta Neurologica Scandinavica*, 74(2):150–155, 1986.

[20] K. Lehnertz and C.E. Elger. Can epileptic seizures be predicted? evidence from nonlinear time series analysis of brain electrical activity. *Phys. Rev. Lett.*, 80:5019–5022, 1998.

[21] P. Loiseau. Incidence et evolution a 1 an des syndromes epileptiques en gironde. *These No. 338, Universite de Bordeaux II*, 1987.

[22] J.C. Sackellares, L.D. Iasemidis, D.-S. Shiau, R.L. Gilmore, and S.N. Roper. Epilepsy - when chaos fails. In K. Lehnertz, J. Arnhold, P. Grassberger, and C.E. Elger, editors, *Chaos in the brain?* World Scientific, 2000.

Chapter 14

TESTING WHETHER A PREDICTION SCHEME IS BETTER THAN GUESS

Mark C.K. Yang
Department of Statistics
University of Florida
yang@stat.ufl.edu

D.-S. Shiau
Department of Neuroscience, University of Florida; VA Medical Center, Gainesville, FL
shiau@epilepsy.health.ufl.edu

J.C. Sackellares
Departments of Neuroscience, Neurology, Pediatrics, and Biomedical Engineering, University of Florida; VA Medical Center, Gainesville, FL
sackellares@mbi.ufl.edu

Abstract To be able to predict a future event is the most convincing side of science. Usually in the beginning of an investigation, the prediction is not perfect, i.e., an event may be missing or a prediction turns out to be a false alarm. When past prediction records are available, can we determine whether the prediction scheme is promising? In this paper, we use the naive optimal prediction as a yard stick, i.e., testing whether the new prediction scheme is better than the naive optimal scheme through statistical hypothesis testing. Here the naive scheme is defined as we know only the distribution of the inter-arrival times without any other auxiliary information. We use the trade off curve between false alarm rate and sensitivity to measure the prediction performance and the bootstrap method to compute the p-value. Real data and simulation examples are presented.

1. Introduction and Modeling

To be able to predict the occurrence of a disease episode has many important clinical applications. The onset of stroke, hear attack, or severe pain are examples. The particular example that motivates this paper is the prediction of seizures from epilepsy patients. Epilepsy is among the most common disorders of the nervous system. The incidence rates range from 24 to 53 per 100,000 per year [3, 6]. For epilepsy patients, the ill interferes with normal life and sometimes fatally only when the a seizure occurs. Thus, any methods that can predict its occurrence would be a great clinical achievement. Unfortunately, there still no method at this moment that can predict seizure with high reliability [7, 9]. Thus, we are seeking valid methods that may eventually lead to a clinically useful prediction. How do we quantify a valid method? In this paper, we propose a method that will show whether a given method is better than a naive method based on some trivial information. In our case, the Lyapunov exponent aspect of EEG is used to predict seizures, but the prediction is far from perfect. Is Lyapunov exponent of EEG really useful in prediction?

To measure prediction accuracy is not an easy matter. An obvious measure is the prediction error defined by the distance between the predicted time and the real occurrence time such as the mean square error used in time series prediction. The top scenario of Fig. 14.1 shows this possibility. We can use either $|y_i - t_i|$, or $|y_i - t_i|^2$ to measure the prediction error. However, this definition does not work for the other two scenarios in Fig. 14.1 . In the middle scenario, there are three events occurred before the predicted time y_1. There seems no way to defined the prediction error by time difference. Also, for the bottom scenario, when the first prediction y_1 failed to predict an occurrence, the predictor cannot wait indefinitely. He will make a new prediction when there are signs for a new event. So he predicts again at y_2, and y_3. It is again difficult to measure the prediction error by time difference.

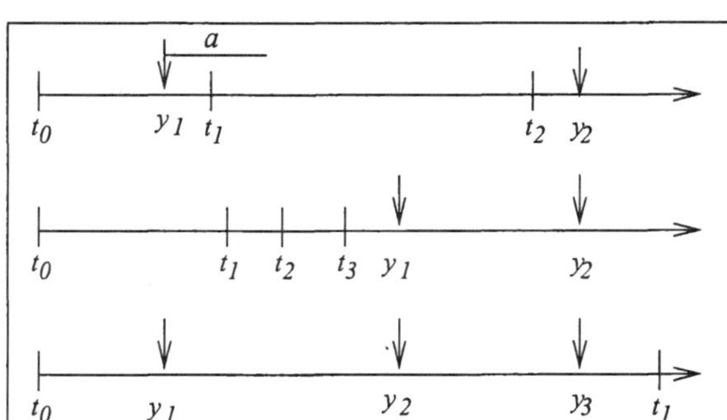

Figure 14.1. Possible scenarios between real events t_0, t_1, etc. and predicted times y_1, y_2, etc. The length a is the length of the alert period.

Thus a commonly used criterion for event prediction is the trade off between sensitivity and false alarm rate [12]. Roughly speaking, sensitivity is defined as the probability of making a correct prediction of an event and the false alarm rate is the number of false alarms per unit time. The unit time can be one minute, one hour or 24 hours or any time interval. This causes some inconvenience because the time unit can be subjective. We define the false alarm rate as

$$\begin{aligned}\phi &= \text{ average false alarm per event} \\ &= \lim_{n\to\infty} \frac{\text{number of false alarms in predicting } n \text{ events}}{n}\end{aligned} \tag{1}$$

Note that this definition is equivalent to false alarm per unit time when we know the mean occurrence time.

To define sensitivity, we cannot avoid one new parameter, the *alert period* (denoted by a in this paper), i.e., if the event occurs within the alert period after the prediction (warning) starts, it is a correct prediction. Otherwise, the prediction is a false alarm. With this new parameter, we define the sensitivity as

$$\psi = \text{Probability an event is correctly predicted.} \tag{2}$$

Ideally when ϕ increases ψ will decrease, but it may not be so if the prediction scheme is "strange". The computation of ϕ and ψ for a general renewal process is given in the Appendix.

For a fixed false alarm rate, we can compare two prediction schemes by their sensitivity, or conversely, compare their false alarm rates at a fixed sensitivity. Unfortunately, it is almost impossible to fix the false alarm rate in a sample with a small number of events. For example, suppose the three figures in Fig. 14.1 are three samples from different patients. The estimated ψs and ϕs are shown in Table 14.1. Thus if we fix the false alarm rate at 1.0, there is no estimates for ψ. Thus we choose to use the the trade-off curve in Fig. 14.2 which is similar to the OC (operating characteristic) curve. We still use the name OC curve for the ϕ-ψ relation. Usually a prediction scheme has many parameters for tuning, such as economical consideration. Thus the predictions in Fig. 14.1 are from one option that allows a moderate false alarm rate. The predictor can reduce the false alarm rate by making less predictions. Hence each sample in Table 14.1 merely represents a point in the OC diagram. By tuning the prediction parameters, we can estimate the entire OC curve. A typical estimated OC curve is shown in Fig. 14.3. Actually, this is one curve from real data to be discussed in the example section. Intuitively, the OC curve should be a monotonic curve. The horizontal part such as section b and the vertical part such as section c in Fig. 14.3 can happen in theoretical curves, but the non-monotonic parts d seems to be an undesirable small sample property.

It is obvious that the prediction scheme S_1 is better than scheme S_2 in Fig. 14.2. The superiority is uniform for any fixed ϕ or ψ. But if there is

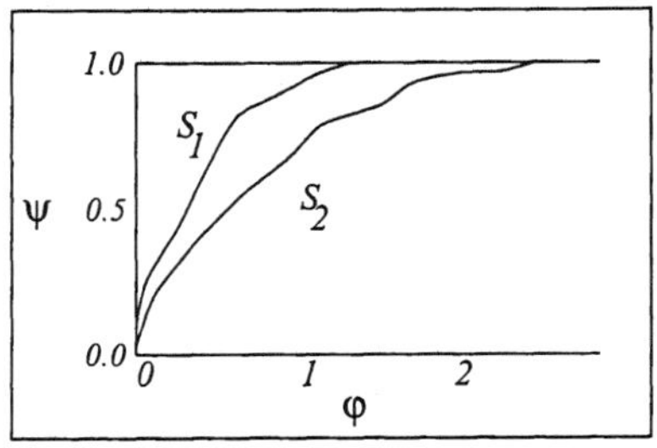

Figure 14.2. Two OC curves by two prediction schemes. The horizontal axis is the false alarm rate/event, and the vertical axis is the sensitivity. S_1 and S_2 are two examples.

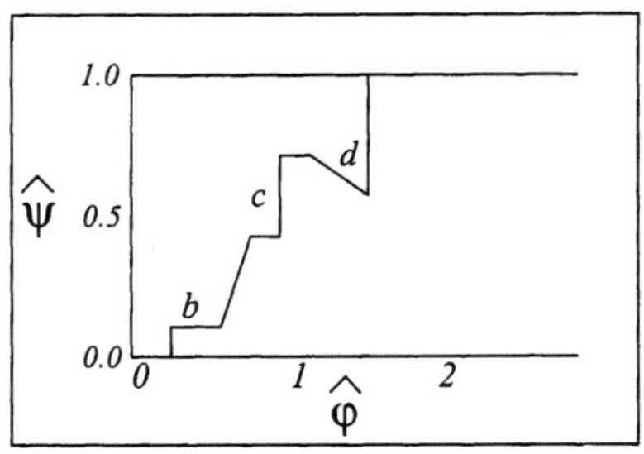

Figure 14.3. An estimated OC curve from a sample.

crossover between the two curves, then it is difficult to define superiority. Actually, there is no reasonable way to define superiority because it depends on which false alarm rate you prefer. But in practice, it would be very difficult to prove the existence or nonexistence of crossover based on estimated curves. We propose to use area above the curve (the upper corner area) A as a measure of the performance of prediction scheme, i.e.,

$$A = \int_0^\infty [1 - \psi(\phi)] d\phi. \tag{3}$$

Note that this area always exists in theory as well as in sample, because if we make a prediction immediate after an event and after each prediction failure, $\psi = 1$ and ϕ is bounded by the mean arrival time and alert period a. Thus, the

Table 14.1. Estimated sensitivity $\hat{\psi}$ and false alarm rate $\hat{\phi}$ of Fig. 14.1.

Estimate	*Top figure*	*Middle figure*	*Bottom figure*
$\hat{\phi}$	0.5	0.0	1.0
$\hat{\phi}$	0.5	2/3	2.0

curve is closed at the right end. Also, if no prediction is made, it is closed at the left end. Hence the area above the curve is well defined. Obviously, the smaller the A, the better is the prediction scheme. Other areas of statistics also use the whole curve as a comparison of different schemes or treatments. It is usually called the area under the curve [1, 2, 11]. However, our OC curve is different from the traditional specificity in the x-axis, because ϕ can be larger than 1. That is why we use the area above the curve, because the best value is the well defined number 0. If we used the area under the curve, the best value would depend on the range of ϕ which can vary.

2. Naive Prediction Schemes

If there are existing prediction schemes, we can compare a new scheme with the best existing one. But when the predict scheme is the first of its kind, we have to justify its merit with some naive or trivial prediction scheme.

Two naive schemes were suggested by [8], the periodic prediction and random prediction. The former means to predict periodically every so often until the next event appears and the latter is to use a random guess for the next event. There is some uncertainty in the second scheme because random guess can be defined in many different ways. We feel that a nature random prediction is to let the next prediction be an exponential random variable with mean μ. When we adjust μ, we will get an OC curve. Similarly, we can obtain the OC curve for the periodic prediction by varying the period μ.

The two measures, sensitivity and false alarm rate, can be combined if there are costs associate with them, i.e., there is a cost c_1 for missing a events and a cost c_2 for one false alarm [12]. In this case, the total cost of a prediction scheme is

$$C = c_1(1 - \psi) + c_2\phi. \tag{4}$$

In general, the optimal scheme depends on the cost and the occurrence pattern. However, under a mild condition, if the event arrivals have a non-decreasing hazard rate, then the optimal prediction is to predict once and keep constant alert thereafter, where the hazard rate is defined as

$$\lambda(t) = \lim_{\Delta t \to 0} \frac{\Pr\{\text{The event will happen in } (t, t + \Delta t) \text{ given it has not occurred yet}\}}{\Delta t}. \tag{5}$$

This scheme is optimal regardless of the cost structure in (4). A proof is given in the Appendix. Since this is a prediction method that required no prior information of the events, we defined it as the optimal naive prediction scheme. To assume non-decreasing hazard rate is quite reasonable for medical events, because an acute disease episode usually occurs when the accumulation of certain undesirable waste or fatigue in a tissue exceeds a threshold. Thus the hazard increases with time.

3. Testing the Hypothesis that the New Method is Better

The method described here can be used to compared any two prediction schemes, but we use a new and the optimal naive prediction methods to illustrate the methodology. Suppose there is a sample from a patient. By tuning the parameters in prediction scheme, we can estimate the OC curves and area above the curve for the new prediction scheme ($\hat{A}_n$) and for that for the optimal naive one ($\hat{A}_o$). If their true areas are and A_n and A_o, then we wish to test the hypothesis:

$$H_0 : A_n = A_o \text{ versus } H_1 : A_n \neq A_o. \quad (6)$$

Although we are merely interested in the one sided alternative $A_n < A_o$, we follow the usual caution to do a two sided test in (6). At this moment, we do not know how to find an exact test for (6). A t-type of test was suggested in [11] assuming that the two measures on the x- and y- axes have a bivariate normal distribution which is irrelevant to our work. We choose the boostrap method, i.e., take random sample from the inter-arrival intervals with replacement and compute $\hat{A}_o$ from this sample. The p-value can be found by the position of $\hat{A}_n$ in the lower quantile of all the $\hat{A}_o$s. More precisely,

$$\text{p-value} = \frac{\text{number of } \hat{A}_o \leq \hat{A}_n \text{ in bootstrap sampling}}{\text{bootstrap size}} \quad (7)$$

A small p-value indicates that it is unlikely that the optimal naive method can reach such a small area ($\hat{A}_n$) attained by the new method.

In many practical case, one sample may not contain enough events to produce a small enough p-value. Suppose there are n sample with p-values $p_1, p_2, \ldots, p_n$. We suggest to use the concept of meta analysis to combine them. Let $z_i = \Phi^{-1}(p_i)$, where $\Phi(x)$ is the standard normal distribution function. The combined p-value to test whether the new scheme is better than the naive becomes

$$\text{p-value} = 2\Phi(z), \quad z = \frac{\sum_{i=1}^{n} z_i}{\sqrt{n}}. \quad (8)$$

The reason for this computation is simple, because under the null hypothesis, the p-value should be uniformly distributed in $(0, 1)$. The inverse z-transform make the z_i a standard normal random variable and so is z. The p-value indicates how extreme the $p_1, p_2, \ldots, p_n$ are, when H_0 is true.

4. An Example

A new method based on the entrainment of the short-term maximum Lyapunov exponent (STL_{max}) has been constructed to predict seizures [4, 5]. It has some successes and some failures. The question is whether this method is on the right direction toward seizure prediction. Is it better than the optimal naive method?

Seven epilepsy patients with various number of seizures were examined. We examined the inter-arrival times from each patients and found that the inter-arrivals could be considered as independent (by sample autocorrelations) with non-decreasing hazard rates (by histogram). Thus, it is fair to compare the STL_{max} method with the naive optimal method. Table 14.2 show the results from each individual and their joint performance. Though a low significance level is not attained for each subject, the overall p-value 0.015 computed by (8) shows that the performance of STL_{max} described in [4, 5] is not an artifact. We also tested this method by comparing the optimal naive method with the periodic and random prediction schemes. They are not significantly different, although both method seems to be on the worse side. If they reached an overall p-values larger than 0.95, we would conclude that they are indeed worse than the optimal naive method at the usual 0.05 significance level testing.

5. Power Analysis

To see how powerful this testing scheme is, we choose to use a compound Poisson process [10] to generate events. Suppose chemicals are cumulated in a tissue and a breakdown (event) occurs when the cumulated chemical excesses a threshold (see Fig. 14.4). Suppose the arrival of each chemical follows a Poisson process. Without loss of generality, we let the chemical added to the tissue to be 1 (unit of chemical) and the mean arrival time is 1 (unit of time). Let the time t be 0 after each event. Thus, the cumulated chemical x_t at time t is k where k is the number of arrivals between the previous event and t. Let the threshold for the event to occur be τ, i.e., an event occurs as soon as $x(t) \geq \tau$.

Suppose the predictor knows this process and the threshold τ, but cannot measure the chemical level precisely. Let the observed chemical level be

$$y_t = x_t + \epsilon_t, \tag{9}$$

where ϵ_t is the noise level, assumed to be normally distributed with 0 mean and variance σ^2. Based on this value, the predictor has to make a decision whether

Table 14.2. p-values of testing the hypothesis that the STL_{max} is better than the optimal naive method. The "Size" row indicates the number of seizures examined. The last two rows are the comparisons of the periodic and random prediction methods with the optimal naive method.

Subject	*CHAP*	*MASH*	*FUS*	*PRE*	*MORG*	*MERR*	*PHEL*	*Overall*
Size	8	15	19	6	8	17	17	90
STL_{max}	0.344	0.104	0.190	0.090	0.336	0.102	0.196	0.015
Periodic	0.344	0.606	0.496	0.332	0.760	0.328	0.378	0.813
Random	0.344	0.794	0.778	0.332	0.452	0.574	0.378	0.848

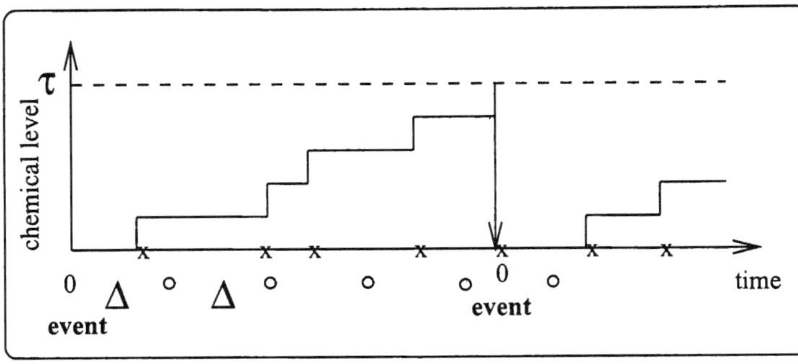

Figure 14.4. Events generated by a compound Poisson process. Chemical of 1 unit arrives at x with a Poisson distribution. An event is triggered when the accumulated level reached τ. The predictor can make an observation (o) at each Δ interval.

to make a prediction and how long ahead if the prediction is to be made. Since the expected occurrence time of the next event is at

$$T = \tau - x_t, \tag{10}$$

we should predict the next even at $\tau - y_t - a/2$, where a is the length of the alert period. This prediction value has variance $\tau - x_t + \sigma^2$ which can be estimated by $\tau - y_t + \sigma^2$ since x_t is not observable. Apparently if $\tau - y_t$ is large, we should not prediction, because the prediction tends to be inaccurate. Whether to predict also depends on when we can take the next measurement. We let this new parameter be Δ, i.e., the chemical will be measured every Δ unit of time (see Fig. 14.4). As mentioned before, the optimal decision should depend on a given false alarm rate, or a given sensitivity, or costs in (4), but these parameters are usually difficult to come by, especially at the research stage. In this example, the main purpose is to see whether the observed chemical level has any prediction value. We use the following prediction rule: to make a prediction if the observed level y_t is higher than y_o and when we predict we set the alter period at

$$\hat{T} = \theta(\tau - y_t - a/2) \tag{11}$$

where y_o and θ are tuning parameters. When $y_o = \theta = 0$, it is the extreme case that we are alert all the time, and $y_o = \infty$ is the other extreme case that we never make any prediction. These two extreme conditions will produce a closed OC curve.

The power is defined as the probability that this chemical measurement is considered as a useful predictor by accepting H_1 in (6) by test (7) when n events can be observed. We expected that power will increase when n increases or σ^2 decreases. However, even there is no measurement error ($\sigma^2 = 0$), the power is not 1.0 because the prediction error variance $\tau - x_t + \sigma^2$ is still not 0.

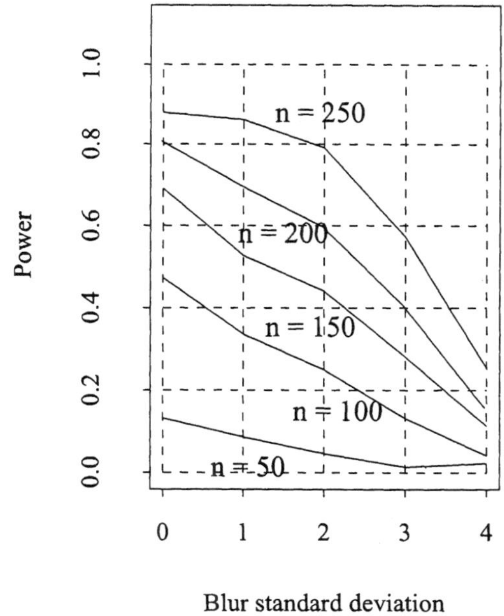

Figure 14.5. Power curve for a scheme with inside information against the optimal naive prediction scheme, where n is the number of events in the sample, Blur standard deviation is the measurement error of the inside information (see Eq. (9)).

There are many parameters that affect the power of this prediction scheme. Only one simulation study is shown here. In this case, the threshold τ is set at 20 (units of time), in other words, the inter-event arrival time has expected value 20 with standard $\sqrt{20} = 4.47$. Thus there is quite a variation on the event occurrence if no additional information is given. The alert period is set at 4 and chemical will be examined at $\Delta = 2$ time units. The tuning parameter y_o is set at 0, 5, 10, and 15 and θ at 0, 0.5, 1.0 and 1.5. The results are given in Fig. 14.5, where n is the event size. The curve shows what we expected; power increases as n increase, and as the blur level σ decreases.

This result shows that the optimal naive scheme is difficult to beat with small sample sizes.

6. Concluding Remarks

1 The key contribution of this research is to provide a rigorous statistical test for validating any predicting scheme. The power of this test is also examined with a reasonable model.

2 For the most commonly used exponential arrival, the naive optimal scheme is either be alert all the time, or give up prediction if there is no inside information.

3 For perfect prediction, the sensitivity reaches 1 with no false alarm. The area about the curve A is 0. This is, of course, very easy to prove its validity except the events occurs periodically with an exact period. In this case the naive optimal scheme also has $A = 0$.

4 During our simulation experiment, we found that the optimal naive scheme can be difficult to beat. For example, if the event arrival is pretty regular, then any auxiliary information help very little in prediction. This fact, obviously agrees with common sense: When a event comes regularly, good prediction scheme are no big deal. Any prediction scheme has to prove its validity by a very strong evidence (large sample).

5 The method developed in this paper does not depend on the cost of the false alarm or missing a event. But when the method is mature enough for practical use, the costs have to take into consideration. At this stage, a test based on given false alarm rate or cost structure should be developed.

Appendix: Optimal Prediction Strategy for the Next Event

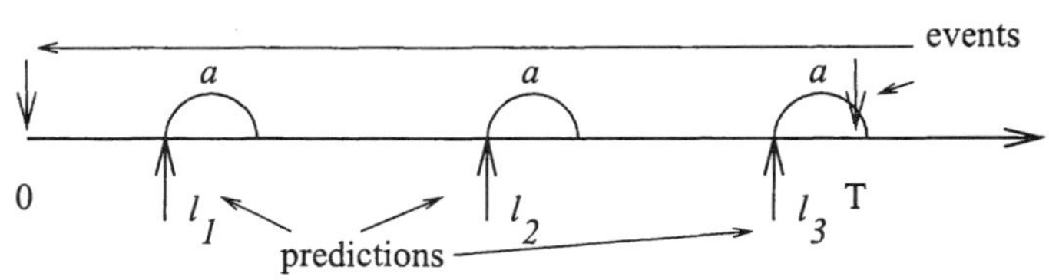

Figure 14.A.1. The relation between predictions l_i, events (down arrows), and alert periods.

The general prediction scheme can be described as Fig. 14.A.1, where prediction occurs at times $l_1, l_2, \ldots$ before the event occurs at T. When there is no other auxiliary information before the the occurrence, the l_i's are pre-determined. Let $I_A(x)$ be the indicator function with value 1 if $x \epsilon A$ and 0 otherwise. Then the number of detection and false alarms are respectively,

$$\begin{aligned} I_\psi &= \sum_{i=1}^{\infty} I_{[l_i, l_i+a]}(X), \\ I_\phi &= \sum_{i=1}^{\infty} I_{[l_i+a, l_{i+1}]}(X), \end{aligned} \tag{14.A.1}$$

where X is the random variable of occurrence time T in Fig. 14.A.1. Taking expectation on (14.A.1), we find the ψ and ϕ defined in (1) and (2) with the following expressions.

$$\begin{aligned} \psi &= \sum_{i=1}^{\infty} \int_{l_i}^{l_i+a} f(x)dx \\ \phi &= -1 + \int_0^{l_1+a} f(x)dx + 2\int_{l_1+a}^{l_2+a} f(x)dx + 3\int_{l_2+a}^{l_3+a} f(x)dx + \cdots, \end{aligned}$$

where $f(x)$ is the density function for X. The optimal allocation of $l_1, l_2, \ldots$ that minimizes the total cost function (6) for general occurrence density function $f(x)$ is not easy to obtain. Some special cases such as X is exponentially distributed can be worked our analytically. A special case is when the hazard function $\lambda(x)$ is monotonically nondecreasing. In this case, we may derive the optimal prediction strategy in the following manner.

Suppose we wish to predict that the event will occur at x given it has not occurred before x. Then the loss is then

$$\frac{-c_1}{1-F(x)}\int_x^{x+a} f(x)dx + c_2[1 - \frac{1}{1-F(x)}\int_x^{x+a} f(x)dx].$$

We should make the prediction is the loss is less than than 0. It is equivalent to

$$\frac{\int_x^{x+a} f(x)dx}{1-F(x)} \geq \frac{c_2}{c_1+c_2}. \qquad (14.A.2)$$

For small a, the first term of (14.A.2) is approximately $a\lambda(x)$. Thus if $\lambda(x)$ is a nondecreasing function, the attendant should be at constant alert as soon as (14.A.2) is satisfied. If this approximation is not good enough, then the condition is if $[F(x+a) - F(x)]/[1 - F(x)]$ is a monotonic increasing function of x.

A similar result was obtained in [12], but the result seems counter intuitive. In it, there are three costs, c_1 = cost of predicting an event, c_2 = cost of a false alarm and α = cost per unit time of maintaining the alert. His result shows that if $\alpha = 0$ we should be alert all the time. This is counter intuitive because with a large overhead cost c_2, we should also avoid false alarms. Our result (14.A.2) are very intuitive for large c_2 (no prediction) or small c_2 (constant alert). Moreover, the maintenance cost during the alert period can be easily embedded in c_2, i.e, a false alarm cost is the sum of a initial preparation cost and a maintenance cost for the alert period. Thus, the alert cost in [12] seems unnecessary.

References

[1] S.B. Cantor and M.W. Kattan. Determining the area under the curve under the roc curve for a binary diagnostic test. *Med Decls making*, 20:488–470, 2000.

[2] D.L. Fairclough. *Design and Analysis of Quality of Life Studies in Clinical Trials*. Chapman & Hall/CRC, Boca Raton, 2002.

[3] W.A. Hauser and L.T. Kurland. The epidemiology of epilepsy in rochester, minnessota, 1935 through 1967. *Epilepsia*, 16:1–66, 1975.

[4] L.D. Iasemidis, P.M. Pardalos, J.C. Sackellares, and D.-S. Shiau. Quadratic binary programming and dynamical system approach to determine the predictability of epileptic seizures. *Journal of Combinatorial Optimization*, 5:9–26, 2001.

[5] L.D. Iasemidis and J.C. Sackellares. Chaos theory and epilepsy. *The Neuroscientist*, 2:118–126, 1996.

[6] P. Joensen. Prevalence, incidence and classification of epilepsy in the faroes. *Acta Neurologica Scandinavica*, 76:150–155, 1986.

[7] K. Lehnertz and C.E. Elger. Can epileptic seizures be predicted? evidence from nonlinear time series analysis of brain electrical activity. *Phys. Rev. Lett.*, 80:5019–5022, 1998.

[8] T. Maiwald. The seizure prediction characteristic: A new terminology and assessment criterion for epileptic seizure prediction methods. Presented at the Conference on Quantitative Neuroscience, University of Florida, February 2003.

[9] H.L.V. Quyen et al. Anticipation of epileptic seizures from standard eeg recordings. *Lancet*, 357:183–188, 2001.

[10] S.M. Ross. *Stochastic Processes*. John Wiley & Sons, New York, 1983.

[11] E.F. Schisterman, D. Faraggi, B. Reiser, and M. Trevisan. Statistical inference for the area under the receiver operating characteristic curve in the presence of random measurement error. *Am J Epidemoil*, 154(2):174–179, 2001.

[12] D. Vere-Jones. Forecasting earthquakes and earthquake risk. *Internal J Forecasting*, 11:530–538, 1995.